RAG TOP

RAG TOP

BY HENRY GREGOR FELSEN

Octane Press, Edition 1.0, February 2026
Random House Edition, 1954
Bantam Edition, 1956
Felsen Ink Edition, 2013

Copyright © 1954 by Henry Gregor Felsen
Copyright © 2026 by Holly Felsen Welch

All rights reserved. With the exception of quoting brief passages for the purposes of review, no part of this publication may be reproduced without prior written permission from the publisher.

ISBN: 978-1-64234-1-348
ePub ISBN: 978-1-64234-2-055
LCCN: 2025941442

On the cover: Cover art used with permission of Bantam Books (Penguin Random House).

Designed by Tom Heffron
Copyedited by Faith Garcia
Proofread by Jody Amato

octanepress.com

Octane Press is based in Austin, Texas.

To Dick Taplinger

CHAPTER 1

The car was a black 1949 Lincoln sedan. There were four people inside, none of them happy. A man, a woman, an adolescent girl, and a younger boy sat in the four corners of the car, each a separate island of discomfort and resentment.

Mile after mile the bulbous, dusty-windowed car scuttled along the highway like a huge, cloudy-eyed beetle hurrying to its hole. With its blunt nose pressed close to the road, the car seemed to be guided toward its goal by its sense of smell, rather than by the hand of the man inside.

But the man was in control, both of the car and the people in it. He and he alone knew where they were going, and why. His wife and children knew only that they had to make the trip with him; that there was no way to avoid the ill-tempered expeditions into the country he insisted on whenever he had an off-duty Sunday.

The man was about forty-five, of medium build, with an angry look of disapproval on his thin lips and in his narrowed eyes. Although it was summer, he wore a dark blue wool suit and a black slouch hat pulled low over his forehead. He sat stiffly erect, his hands in the ten-two position, his head tilted back so he could see over the top of the wheel, like an old farmer driving his first car to church. From this position, he cursed or jeered at all traffic, directing his words at his family, as though they were responsible for all the fool driving that plagued him every mile of the way.

Yet he was no clown, no mere flabby-muscled family tyrant who got his thrills shouting insults at bigger men in newer cars. His face bore the lumps and scars and puckers of battle joined in back alleys, in taverns, on the streets. Marks and dents left on his lean, bitter face and hard body by the fists and nails and teeth and weapons of people who had fought arrest until he had clubbed them into submission.

The knuckles that gripped the steering wheel had been broken so many times they were shapeless. The jagged scar on his left cheek was the souvenir of a broken beer bottle that had been thrust at his eyes. The little patch of bare skin on the back of his head was monument to a hatchet blow suffered when he'd been attacked by a lunatic. His nose had been broken as many times as his knuckles.

Virgil Kern, policeman. A man who patrolled his city from dusk to dawn to protect its peace, he had few illusions about human dignity at any level. On any given night, his tour of duty was a reconnaissance of the vice, brutality, lust, criminality, folly, and depravity of which his neighbors were guilty. And what he saw in the glow of his flashlight stayed with him when the sun was up and sent him desperately probing the country roads on his day off.

In Virgil's mind, the city itself was to blame, because it was big. He had come from a small town and never lived in a big city until the war, when he'd worked in a defense plant. From there he'd gone on to become a policeman. He had come to Des Moines with an established distrust of any community that was too big to know everyone in it. What he had seen as a policeman confirmed his distrust and gave birth to his fears.

The city was too big. There were too many strangers, too many houses, too many streets and alleys and holes where humans could hide. And that's when people got mean, when they thought they had a place to get out of sight, or weren't known. That's when they got wild and dirty.

Kern's eyes flicked in the direction of the rearview mirror. His sixteen-year-old daughter, Darlene, was slumped in the back seat, a sullen expression on her face. She was wearing plaid slacks and a white sweater with short sleeves. She had smoky blue eyes, long blonde hair that she wore in a ponytail, white, smooth skin, and a full figure. He knew she was mad because she had planned to spend the day at a drag race with her crowd of friends. And looking at her, he felt that by having her with him he had managed to keep her safe for another day.

Kern squinted down the road, his hands tight on the wheel. There had to be a place, and he had to find it soon, before it was too late. He had good kids, but they wouldn't stay good in the city. Not from what he knew of the city. Darlene was already running with a wild crowd, where the boys were too old and too rough. He knew the kind. He'd run enough of them down in their hot rods. He knew where they went and what they did. And Doyle. A couple more years and he'd be running the streets and alleys, getting in trouble because the city was big, and there were a lot of places in it to hide.

They didn't know. They couldn't see what they were heading for. But he knew. Night after night he saw what happened to other kids. It was the city. Too big, too dark, with too many places to hide. With too many people to know. Even Agnes didn't understand, and there was no way to tell her. When he tried, she thought he was just talking and wouldn't listen. She didn't think anything could happen to her kids. She didn't know. She hadn't seen what he saw all the time. And it could happen to his kids. Only it wouldn't, because he wasn't going to let it.

And now he was looking again, hoping maybe this would be the day. And afraid it wouldn't be. Afraid there wasn't any such place like the one he was looking for. Afraid of what would happen if he never found it. And his family mad because they didn't know why they were on the road, and he didn't know how to tell them. Because he couldn't tell them that he was angrier than any of them, and at them.

A car came up from behind, honked, and swept past. Kern's eyes narrowed. He was doing sixty. He always drove sixty on the open road. He believed it was the maximum safe speed and that no one else should exceed it.

"I got a good mind to catch that reckless fool and give him a ticket," Kern growled. "Teach him to speed." He looked hard at his wife. He wanted an answer.

Agnes Kern was a tall, bony woman with sunken cheeks, thin lips that drooped at the corners, and straight, graying brown hair pulled back tightly over her head and rolled in a bun at the back of her neck. She sat with her hands in her lap, staring straight ahead, enduring the ride. The sun beating on her through the window made her feel sick and dizzy, but she was afraid to open it. Virgil had told her time and again that opening the windows cut down on the gas mileage, and besides, he didn't want the wind blowing on his neck.

"Agnes?"

"Why Virgil," she said in a thin, meandering voice, "there isn't any speed limit on the highway. Is there, children?"

"That's right, Ma," Darlene said spitefully. "People can drive as fast as they want to."

She would have said anything to contradict her father. She hated him for making her go along on this miserable ride. She had planned to go to the drag races with her crowd, and now all the other kids were at the races, having a good time, and she had to be here, taking another awful Sunday drive with her father, along with her mother and Doyle.

Doyle sat up straight, aroused by the sounds of conflict. He was a thin boy of ten, dressed in faded denims and a sport shirt. He was dark, like his father, with dark eyes and a little furtive grin on his lips. He had been pretending to be asleep while stealthily trying to dirty Darlene's white sandals with his shoes.

"Let's go faster, Pop," Doyle urged, crowding against Kern's back so he could look through the windshield. "Let's race him."

"There is too a speed limit," Kern said angrily, his voice rising. "You look at the Iowa law. It says the speed has to be reasonable and proper. And anybody driving over sixty miles an hour on these narrow roads ain't reasonable or proper. They can be arrested for reckless driving."

"You can't arrest them, Pop," Doyle said provocatively. "We're out of the city limits. You can't arrest anybody outside of Des Moines."

Darlene drew back her foot and kicked Doyle. He kicked back, blindly, still keeping the innocent expression on his face. He didn't mind hearing his father rave. Sometimes when he got real mad he drove faster.

"That's how much you know," Kern said with a short, mirthless laugh. "Anybody can make a citizen's arrest. And they can make it anywhere. All you have to do is see somebody breaking the law, and you can arrest them. I can arrest anybody I please, Doyle. And anywhere I please."

"Would you, Pop?" Doyle asked eagerly.

"Would I what? And git off the back of my neck when I'm driving."

"Pinch somebody," Doyle said. "Why don't ya, Pop? Next guy we see. Will ya, Pop? I never seen you arrest nobody."

"Leave your Pa alone," Mrs. Kern ordered in her tired voice. "We're out for pleasure drivin', not lookin' for people to arrest.

Give your father a chance to enjoy himself and forget his work. And watch your language, Doyle. It's, 'I have never seen you arrest nobody,' ain . . . isn't it, Darlene?"

"No, Ma," Darlene said tiredly. "You used a double negative."

"I did? I didn't know that. What is a double negative, dear?"

"Something you learn in school, Ma."

"You could tell your mother."

"I want to forget school," Darlene said, stretching lazily.

"It wouldn't hurt you to tell me what I said wrong. I don't want to make the same mistake in public. You'd be ashamed of me if I did that."

"Some other time, Ma."

"Now, Darlene—"

"Oh, quit your scrabblin'!" Kern shouted. "You're both drivin' me nuts. If you can't find anything better to talk about, shut up, the both of you."

"I just asked her . . ." Mrs. Kern began to defend her position.

"Don't ask her nothing. You know you won't get a civil answer. 'Less maybe it's some question about boys. That's all she's interested in. Boys. And the crummiest lot of boys in the city of Des Moines, Iowa, if you ask me. There ain't one of them has a thought in his head that ain't to do with rattin' around in some old car, gettin' in trouble with the law."

Mrs. Kern leaned back, her lips tight and prim. "That's enough of that kind of talk," she said.

Kern didn't answer. He directed his fury toward a convertible they had caught up with. In it was a family of five. They were going along about forty miles an hour, laughing and pointing at the countryside.

"I know what they're doing," Mrs. Kern said, her voice sounding bright for the first time. "They're playing that game where you see who's first to see a white horse. Look at that little fella. He's gonna cry because he didn't see it first."

"Pokin' along at forty miles an hour on the open highway!" Kern snarled. "I could arrest that silly outfit for obstructin' traffic. It's just as dangerous to drive too slow as it is too fast. I got a good mind to haul them in."

"One's too fast and the other's too slow," Mrs. Kern sighed. "Don't anybody drive to suit you?"

"I do," Kern muttered. "I drive sixty. That's the right speed for this road. They all ought to be driving sixty."

"Pass 'em, Pop," Doyle urged. "Blast 'em with the horn."

Kern pressed down on the gas pedal. His old Lincoln roared as he swung into the left lane and thundered past the convertible. The children in the back seat waved. Kern leaned over toward his wife and yelled at her closed window. "Git movin' there! What's wrong with you people!"

He cut in viciously in front of the convertible. The other driver tooted lightly.

"Toot at me, will he?" Kern raged. "I'll show that smart aleck! I'll run him in."

"And you wonder why we don't want to go drivin' with you," Mrs. Kern said. "Can't you forget about bein' mad at ever'body and have a good time, like all the others? It's no fun for us, comin' this way."

"We're not out for fun," Kern said darkly.

"Then what . . . ?"

"I'll tell you when I'm a mind to. Now shut up and leave me alone."

"All right, Virgil. If you say so." She sank back into her resigned slump, hands folded, eyes sad and vacant. Then a little light came into her eyes, and she turned toward the back seat. Doyle was sprawled in his corner, reading a comic book. Darlene had her back on the seat, her legs stuck out and resting on the back of the front seat, near her mother's head.

"Children . . ." Mrs. Kern tried to look past Darlene's feet to see her face. "Children . . ."

"What, Ma?" Darlene answered. She was filing her nails, grimacing each time the car lurched and spoiled her efforts.

Mrs. Kern cocked her head to one side, looking hopefully at the two children. "Let's play the white horse game."

"That's for little kids, Ma," Darlene grumbled.

"Doyle?" his mother asked.

"Huh?"

"Would you like to play the white horse game with your mother?"

"Uh-uh. I wanna read my comic."

Mrs. Kern stared at them, disappointed, yet not knowing how to break through their disinterest. She turned slowly, her eyes resting on the grim profile of her husband. Maybe Virgil . . . ? She giggled inwardly at the idea. It wouldn't hurt to ask. Just wait until there weren't any other cars and he didn't have any drivers to cuss at.

She acted suddenly, on impulse. She raised her arm and pointed to Virgil's side of the car, almost knocking off his hat. "White horse!" she cried.

"Watch out!" Kern shouted. "You trying to wreck us?"

"I saw it first," Mrs. Kern said with some dignity.

"Saw what?"

"The white horse. And it was on your side, too." She looked at him hopefully, pleading inside that he would understand.

"What about the white horse?" he asked, giving her an angry look.

"I saw it. It's a game. You know . . ."

"Saw it!" Kern said in disgust. "Is that any reason to poke my eyes out and near wreck us? You saw a white horse."

The children were giggling in the back seat. Mrs. Kern felt her face turn red. "We used to play it when we were kids," she said. "It was fun." She added wistfully, hopelessly, "I thought maybe we could have some fun like those other people."

"You don't even have the game right," Kern snorted, angry and sad that his wife never got things right. It made him hate her and pity her. "The way the game goes, if you see a white horse you spit in your hand and stamp it, but you can't say you seen it until you see a man with a beard. Then it brings you good luck. I used to play it all the time when I was a kid in Oklahoma. A lot of luck it brought me!" He guided the Lincoln around a curve, ready to brake if they should come suddenly on a farmer poking along on a tractor. "You don't see any more horses much," he said resentfully. "White or any other color. And if you do, how long do you think you'd have to wait to find anybody with a beard these days? Nobody wears beards any more. How can you play that game?" He expelled air from his nostrils in a short explosion of disgust. "Anyway, I told you we ain't lookin' for fun."

Mrs. Kern stared out her window, praying hopelessly that they would suddenly pass a man with a beard.

"What are we lookin' for, Pop?" Doyle piped up.

"I'll tell you when I find it," Kern said mysteriously. He smiled grimly to himself. "I know what we're lookin' for. And I'm gonna find it."

"Even so," Mrs. Kern said, half to herself. "On these rides we could have fun like other people if we tried."

A wrangle flared in the back seat. Darlene, furious, was trying to slap Doyle, accusing him of dirtying her white shoes. He was protecting himself with his feet and his comic book, screeching half with laughter and half in alarm.

"Children . . . children . . . behave," Mrs. Kern said tiredly.

". . . gonna stop this car and beat the daylights outa the both of you, so help me!" Kern's bellow of wrath brought a momentary cessation of hostilities, but both Darlene and Doyle tried to explain their side at the same time. Their voices mingled shrilly.

"Stop it!" Kern roared. "Stop it! Sit apart back there and shut up, or I'll . . . I'll . . ." He pounded the steering wheel in nervous rage.

". . . can't have fun like other people . . ." Mrs. Kern said forlornly to herself.

"Oh..." Kern choked, his scarred face twisted in helpless anger. The big black car rolled on smoothly, at exactly sixty miles an hour.

CHAPTER 2

"Git your feet down," Kern said, looking at Darlene in the rearview mirror. "We're comin' into town."

"I've got slacks on."

"I said..."

"Oh, all right." Darlene let her feet slip from their perch beside her mother's head and thump to the floor. She sat up primly, her hands in her lap, knees together. She was a very pretty girl, with a curious, friendly face and doll-like features. She looked very clean, very young, and very innocent. At sixteen, she had the figure of a young woman. Kern looked at her in the mirror, noticing the childlike face.

They had come around a curve at the top of a hill, and the road dropped down to a bridge and straightened out to lead them into the next town.

"Dellville," Doyle said, reading the sign. "Speed limit twenty-five miles an hour. You're doing thirty, Pop."

Kern ignored Doyle. He was examining the town. He looked at the side streets, the houses, the trees. He drove around the town square, noticing the benches, the band shell, the number of stores that were open, the number of store buildings vacant. Then he drove slowly through the various side streets, silent and watchful. When he had seen all of the town, he returned to the square and parked head in in front of the drug store. He turned off the engine, pushed his hat back on his head, and began to roll a cigarette.

"How do you like it?" He licked the cigarette tight, put it in his mouth and lit it with a kitchen match, consuming a third of the cigarette in the initial blaze.

"Pleasant," Mrs. Kern said. It was what she always said when Virgil asked her opinion of a new town.

Darlene looked at the deserted street, the few cars, the old men in the shade of the trees in the square. "It's so dull."

Doyle squirmed restlessly in the back seat. "I'm hot," he complained. "Let's get us a soda."

"You'll git one. Don't rush me."

They sat in the car, the children bored and uncomfortable, Mrs. Kern tired, wishing she were home, Virgil alert and watchful as he scanned the town from under his black hat. Suddenly he poked his wife in the ribs with his elbow. "There," Virgil said, his lean face scornful. "There he is. How'd he ever git the job?"

"He" was a policeman. He had come out of a little red brick building and was standing on the sidewalk, chewing on a toothpick. He was an old man with a huge belly, wearing shiny blue pants and a gray shirt. His gun belt, passing under the swelling belly, looked as though it was hardly above his knees. The old man had a large, fat face, with so many bulges it looked as though his skin had been put on over melon halves. A small blue police cap perched squarely atop his big head, like a squashed cupola atop a square Victorian mansion. He wore tiny, rimless glasses.

"Policeman," Virgil muttered derisively. "I'd like to see him in my job just one night. There ought to be a law against people like that being the police. I can imagine how scared anybody'd be of him."

"Well," Mrs. Kern said, changing her position slightly. "We've seen him."

"Ain't we gonna have a soda before we go?" Doyle wailed.

According to the pattern, it was time for Virgil to start the car and drive home, silent and seemingly disappointed. It was always like that. On his day off, he always took them for a drive and insisted they go along. He seemed to know where he was going, although he never told any of them in advance. He just drove until he came to some town. Then he would drive all over town as he had done here. Sometimes he went on to another town, and sometimes, as on this day, he parked in the center of town and waited. Waited until he had seen the policeman. Then he would drive home. It was meaningless to his family, but he seemed to have some purpose in mind. At least he did before they started out. On the way home he always acted as though he had lost something or had something stolen from him. They couldn't understand why he made the trips, and they were afraid to ask.

Kern reached in his pocket and drew out some change. He handed the money back to Darlene. "You and Doyle git a soda," he said. "Don't take all day." He looked at his wife. "You want one, Agnes?"

"We might all go in and have one," she said. "That would be nice."

"I'll wait here," Kern said. "There's something I want to see." He was staring at the fat policeman.

"I'll wait with you," Mrs. Kern said. "Go on in, children. And don't start reading comics and forget to come out."

The children went into the drugstore, Doyle bounding ahead, Darlene following.

"She's pretty," Mrs. Kern sighed.

"Too pretty," Kern said. "For no more sense than she's got."

"She's not a dumb girl, Virgil. She's smart, when she wants to be."

"If she's so smart, how come she hangs around with that motorcycle crowd? They're a tough bunch."

"Just noisy," Mrs. Kern said. "They come to the house. They're just neighborhood boys she's growed up with."

"Let her run wild," Kern said. "You know what'll happen, don't you?"

"That's a way to talk about your own daughter!"

"My daughter ain't no different from anybody else's daughter," Kern said angrily. "I see 'em. You don't. Maybe I ought to take you on my rounds some night. Don't tell me about what can happen to daughters."

"You can't lock her in the house the rest of her life," Agnes Kern said.

"I can git her away from them punks she hangs around with. See to it she meets some decent people."

"And just where do you . . ."

"Here," Kern said, his eyes on the fat policeman.

"Us move here?" She looked at the town with new eyes.

"I growed up in a small town," Virgil said. "It's the only place in the world to raise kids. Where things is quiet and decent. I seen too much of the city, Agnes. Seen too much of what it can do to girls like Darlene and boys like Doyle. I aim to git 'em out."

"They wouldn't want to leave their friends, Virgil. The children wouldn't like . . ."

He began building another cigarette. "I don't give a hoot what the children like," he mocked. "I'm runnin' this family, and I'll run it my way. I know what's best for 'em. All they want's excitement and trouble. We git 'em down here, and there's a chance they'll grow up to be decent. Besides," he added in a lower tone, "you ain't gonna say a word to the children. I ain't sure we'll be comin' here yet. They can find out we're movin' when they see the furniture goin' out. I don't want to give that girl any excuse for any last flings, if you know what I mean."

"I don't know what you mean about anything," Mrs. Kern said wearily.

"I been lookin' for a town like this," Kern said. "Some nice, decent place. Like the town I growed up in. And one that would need a good policeman. Look at that fat old slob. Bet this town'll be glad to git rid of him and git a real policeman."

"You, Virgil?"

"Me." He exhaled a thin cloud of smoke through his nostrils. "I've found me a good town, I think. I looked long enough, but I found it. And I'll see that it stays good." A note of hoarse pride came into his voice. "That's the kind of father I am. I do something to protect my kids. You go in and have a soda with 'em, Agnes. I'm gonna find the mayor and have a talk with him. But don't you dare say a word about this to the kids. Understand?"

She got out of the car and stood by the door. He got out, stretched, and adjusted his hat. "What are you waitin' for?" he demanded, seeing her standing by the car, looking at him with her tired, submissive eyes.

A thin smile puckered her lips. Suddenly she wanted to laugh, it was so awful. "You didn't give me any soda money like you did the other kids," she said solemnly.

Virgil stared at her through the smoke of his cigarette. She stood before him with her head bowed, the bun riding the high curve of her long neck. She looked meek, beaten. Only in her eyes and at the corners of her mouth was there a hint of the strange laughter that she felt inside. And the fact that she could laugh now, at this, made her feel superior to the scowling man who, disconcerted by her faint smile, counted out the exact change she would need for a soda.

"Thank you," she said gravely, taking the money.

He saw the amusement in her eyes. "Agnes," he said in an irritated tone, "I don't know. For a grown woman, mother of two kids . . . the way you act . . ." He turned and strode away, in the direction of the fat policeman. He walked straight, with his head up and his shoulders back. He marched, she thought. Always marched. Couldn't walk like other people. Had to march. Had to feel the club in his hand even when it wasn't there. If he'd only laughed with her about the money. But he hadn't. He didn't understand laughing. He was afraid of it. The only time he laughed was to make fun of other people. That's what he thought laughing was for. Not to have fun, to make fun.

She watched his tense, aggressive walk and shook her head. Poor Virgil. She didn't have any more fun than he did, but at least she knew there was such a thing. She closed her fingers tightly over the soda money he had given her. Poor Virgil. If he wasn't so hard to live with, she'd feel sorry for him. Now it was this idea of moving. It was always something. Being married to Virgil was like living in a small cage and being taken care of by a bad boy with a big stick.

The family was such a worry to Virgil! He was always afraid something bad was going to happen to them. Afraid Darlene would get in trouble with boys or that Doyle would get mixed up in some mischief. Sometimes she had the feeling that Virgil would be happy if he could just lock them all up safe in jail. Sometimes she felt he wanted to lock up everybody in the world. Locking people up was his answer to everything. The only way he knew of keeping people from doing harm or being harmed. He'd been her jailer more than he'd been her husband. Now he wanted to lock up the children in this little town. Poor Virgil. It would be so much easier to stand up against his meanness and stubbornness if he didn't mean good by what he was doing. But he meant well, and that was what made it so awful. That's why you couldn't fight back. It was like fighting on the side of sin. How did some people do it? How did some people manage to talk and laugh?

She looked up and saw the old men sitting on the benches in the shade. One of the old men had a beard. She looked around quickly, furtively. No one was watching. She pretended to cough, spat in her palm, stamped it and made her wish. Then she went into the drugstore to have her soda.

Doyle was reading an adventure comic; Darlene was looking at one about love. They started guiltily when they saw her and hurried to put the books away.

"There's no hurry, children," Mrs. Kern said. "Your father has some business to attend to. He'll come for us when he wants us."

Darlene reached for her book again. Her hand stopped in mid-air as she heard a familiar, exciting sound. It went by bubbling throatily. The sound of a car with a dual exhaust manifold, cruising slowly. Darlene glanced at her mother, saw she wasn't looking, and stepped quickly to the door, her blue eyes wide and interested. She looked out just in time to see a yellow Chevy convertible rumble past. She tried to see the driver, but his head was turned away. All she could tell was that he had black hair and seemed young. She sighed at having missed his face. He might have been cute.

CHAPTER 3

AT EXACTLY TWENTY MINUTES AFTER THREE, just as he did on every hot afternoon, Arnold VanZuuk steered his black police car toward the corner of Eighth and Oak. The shade would be right on Oak Street, and he liked to sit there for fifteen minutes and think.

Only this day was different.

This day, the heavy-bellied old man who had been Dellville's policeman for thirty-two years was not alone. Riding beside him was the stranger that the Dellville City Council had found and hired to be Arnie's night man. They had told him about it after. Not asked him, told him. As if he needed a night man. He bunked in the jail if he was needed, and he knew which nights to stay up, and what for. After football games, after basketball games, and on Saturday nights. He even knew ahead of time who would go home peaceably and who would spend the night with him at the jail.

Dellville was a good town. In all his years as its policeman, he couldn't remember more than two or three serious troubles. Still, he knew some people weren't satisfied with him. The way the council had been talking, he knew something was in the wind. Speeches about changing Dellville from a "backward small town" to a "bustling, modern little city." He knew what that meant. Already there had been suggestions that the city take away the benches from the town square and that the trees should come down to make more parking spaces. And it was no secret from him that they were ashamed because their policeman was an old Dutchman with a big belly.

What Dellville needed, they said, was a modern police force with modern methods to see that the laws were obeyed. A tough force that would protect the decent people.

Now they had it. The tough, modern policeman, with the flat belly and the hard face.

Virgil Kern, his name was. He came to Dellville from Des Moines, where he had been a city policeman. A mean-eyed man in a rumpled black suit who looked on Dellville more as a hostile invader than as its new protector.

"We go now to Eighth and Oak," Arnie said aloud. "For a little breather in the shade. In the winter, I always go there at noon. The boys drive down Eighth Street when they leave the high school. I park where they can see me good." He chuckled deep in his throat. "Then they slow down, and it makes it safe for the little ones to cross the street."

Arnie stopped the car in the usual shade and turned off the engine. When he was alone, he always loosened his belt and pushed his gun out of the way so it wouldn't poke into his stomach when he relaxed. But with Kern in the car, he felt it would be unwise to show such informality. This type of man would take it as a sign of weakness.

"It's not lively here like Des Moines, eh?" Arnie said, leaning back. "Maybe you miss the city after a while?"

Kern sucked at his cigarette, narrowing his eyes. "I like it quiet. That's why I came here."

"It's a good town," Arnie said, feeling easier in his mind. "On Saturday night, you will have a fight to break up in front of Corbey's place, on the east side of the square. It ain't nothing serious. Every Saturday night the Wakefields and the Shaffers go there for a few beers. They got an old feud, them two families. They don't even remember what it started over. But around eleven o'clock the fight starts. So go there a little before eleven, let them land a few punches, and send them home."

Arnie's big belly shook with his laughter. "I been breakin' up that fight so long now, them fellers wouldn't know how to finish if I didn't show up to stop 'em. Ya . . . ya . . . you'll see, Kern. It's a good town."

Kern detected a faint, pleading note in the old man's voice. The right side of Kern's mouth twitched, but he didn't answer. He got out his tobacco and papers and built a cigarette, lighting it with a kitchen match. He blew out the match, snapped it between his fingers and dropped it on the floor of the police car. There was the old fool's answer if he had any idea of making a nursemaid out of his new cop.

Arnie frowned. His Dutch sense of cleanliness was offended by the spilled tobacco and the match thrown on the floor of the car. He would take that up later with Kern. When he knew the man better.

"You have much trouble with the young punks?" Kern asked.

"Well, now," Arnie said slowly, reprovingly, "we ain't got here what you call punks. We got boys. Some lively, some quiet, some a little naughty. But no punks, Officer Kern. No punks."

He didn't have to get huffy about it, Kern thought. But he was pleased by the old slob's answer. No punks. He could be easier in his mind now about Darlene. He wouldn't have to be scared of every parked car he investigated. It would be easier to watch her here.

And the boys would be nicer. Here in Dellville, he'd see to it that she got in with the nice kids from the nice homes. The ones who had some decent bringing up and were polite. And he could watch her here, protect her until she got married. Like a father should.

He was relaxing, beginning to pat himself on the back for having found Dellville. It had taken a long time, and a lot of driving to a lot of towns, but he'd found it. The safe little town with the old cop, at a time when the town was thinking of hiring a new cop. Just relaxing . . .

Kern stiffened when he heard the sound. He listened, his mouth open a little, his fists clenched.

"It's nothing," Arnie said, noticing Kern's sudden tension. "Just one of our boys with straight pipes. Sounds like Link Aller."

A yellow Chevrolet convertible swung into view at the head of Eighth Street and turned in the direction of the two policemen. Although it was moving slowly, the deep bubbling sound of the twin exhaust pipes rolled ahead of it like a surf.

Arnie chuckled. "Watch this boy," he said to Kern. "He's a pistol, that one."

VanZuuk watched with an amused expression on his broad, shiny face as the yellow convertible rolled down the street well within the speed limit. A block away, the driver spotted the police car as Arnie knew he would. A moment later the mellow tone of the mufflers was swallowed up by an angry, earsplitting roar as the yellow car shot ahead with tires screeching, leaving black marks on the pavement. As the car tore past the police car, the driver turned his head and gave the two policemen a bold, defiant stare. VanZuuk laughed and waved at the black-haired boy who was bent over the steering wheel.

Kern leaned forward with a sudden movement, as though he were going to jump through the windshield and take off at a dead run after the Chevy. He stared in anger and disbelief at VanZuuk, who had made no move to give chase with the police car.

Kern was so angry it was hard for him to speak. "Ain't you . . . going to . . . get that punk?"

VanZuuk shook his head. "Nah, nah. Look down the street. He's going slow again."

"He done it on purpose," Kern muttered. "You let him get away with stuff like that?" He pressed his lips together hard.

Arnie laughed. "I let him get away with a little, so he never tries to get away with much. He likes to show me he ain't one that can be bossed. Other kids, they speed when I'm away, and they slow down when they see me. Link waits till he sees me, then he has to speed. Just to show me he ain't afraid of me. Then he slows down by himself, to show he's going slow because he wants to, not because I'm making him do it."

"I ain't used to taking guff from that kind," Kern said slowly. "And I don't rightly care whether that punk likes to be bossed or not. If he don't toe the line when I'm on duty, him and me is apt to tangle. And I ain't much fun to tangle with."

"Ah, Link is all right," Arnie said. "I know the boy. Last year he was in a race with a friend, and the friend and a girl was killed. Seems everybody around here blamed Link for starting the race. Anyway, that's what he thinks. And he feels bad about it, you understand? He's ashamed to look at the parents of the other boy. And the friends. There ain't no way he can say he's sorry, and nobody he can say it to. So, he feels bad. But he feels it wasn't his fault too, you understand. So he's mad because he's blamed for the accident. So, he's still a boy, you understand. How does he show them bad and mad feelings? A chip on his shoulder. Like a kid will when he thinks he's in the wrong and don't want to admit it. You . . . understand? He speeds for a minute in front of me, it makes him feel better, and he slows down. Soon it's all forgotten. He finds a girl . . ." Arnie chuckled. "I seen it happen many times, my friend. We give nature and time a chance and it saves a lot of wear on the policeman."

A girl, VanZuuk had said, and in Kern's mind it came out Darlene. He was silent and grim as Arnie drove back to the police station. One moment he felt hatred and the next despair. This kid in the convertible . . . This was what he had come to Dellville to get away from! This was the kind Darlene went for and that went for her. The reckless kind. This punk stuck out like a sore thumb in Dellville. It wouldn't be any time at all until he smelled out Darlene and she'd be going with him.

The old feeling was in him, and the bitter taste was in his mouth. The taste of fear he'd felt in Des Moines every time he'd stopped to investigate a parked car. The fear that the girl would be Darlene. The fear that had made him take his family from the city, where there were too many places for the kids to hide, and bring Darlene here, where it was safe. Safe! The first day, and he'd found a punk. Now where could he go to keep his daughter safe?

The answer was simple. He wasn't going anywhere. In Des Moines there had been too many boys, too many places. But in Dellville there was only one punk to take care of. The Kerns would stay in Dellville. But this kid, Link Aller, let him come near Darlene and his days in Dellville were numbered.

"Ya," Arnie said as he parked in front of the city hall. "You'll see we got a pretty nice town here, and nice people. Mainly your job will be to find lost dogs. Always look first in the yard by the grammar school. Even in the summer the dogs go there. The dogs, they don't see their kid, you understand, they think right away he's in school, and they run there."

The two men got out of the car. Arnie grunted with the effort it took to get out from under the wheel and on his feet. Kern's mouth twitched. What a policeman! No wonder Dellville wanted fresh blood.

"You can go home now," Arnie said, taking off his cap and wiping the sweat from his bald head with his palm. "You start tonight, eight o'clock. Later if you can't make it. It ain't the big city, you understand? Take things easy, and get to know the people. That makes the work easy, knowing the people. We got nice people in Dellville."

VanZuuk went inside city hall and dropped heavily into the old swivel chair at his desk. He was worried. That rattlesnake-face Kern was too anxious to get tough, too anxious to push somebody around. Arnie rubbed the side of his head angrily. By golly, if that Kern roughed up any of his people he'd fire him right away, council or no council. Dellville didn't need no tough cops. It was a nice town.

Kern remained standing in front of city hall, watching VanZuuk waddle into the building. Then he thought of the impudent kid in the convertible. That was the kid to start with. Whip him into line, and the others would learn some respect for the law. That was the only way to do it. Pick the leader, the toughest one, and cut him down to size. That was enough for the others, usually. It was time these kids learned that the day was over when the law in Dellville was a big fat joke.

CHAPTER 4

Link Aller had mounted a yard light on the front of the old garage behind the house, and in the evening, when he was through driving the delivery truck for Vernham's Market, he parked his convertible under the light and worked on the engine whether it needed work or not. Just to have something to do.

The hood was up on the convertible, and he sat a foot or so away on a wooden box, cleaning already-clean spark plugs. The bugs were thick around the light over his head, and once in a while a mosquito dropped down to make a pass at him, but Link waved the pest away without being aware of its attack and his defense.

Link was eighteen but had no trouble passing for twenty-one when he wanted to go into some strange tavern and get a beer. He had straight black hair cut short, a long narrow head, and a narrow face. His eyes were black, his nose long and slightly curved. A shave lasted him for only a few hours, after which his beard began to show and made old-looking hollows and shadows on his narrow face. His tight mouth and slanting chin added to the taut, aggressive line of his features. He was slender, but wiry and muscular, like some scantily fed hunting animal that never ate enough to sustain a full life but was hard to kill. He wore dungaree boots, old Levis, a black leather jacket, and a soiled khaki cap.

In contrast to Link's worn, soiled clothing and the junk-littered yard, the yellow car was spotlessly clean. The body was polished and rubbed to a high gloss, the chrome trim sparkled, and there wasn't a vagrant drop of oil or grease on the shining engine.

Most nights he was able to sit under the bright harsh light for hours, working, killing time, and having fun just being with his car. But there were other nights, and this was one of them, when he couldn't lose himself in his work. It scared him when he lost interest in working on the rag top. If he didn't have that, what was left?

He sat on the box, toying listlessly with the spark plug. He didn't need to clean the damn thing, and didn't want to, but there wasn't anything else he wanted to do. The idea of driving around Dellville bored him, and it wasn't any fun to drive up to Des Moines and back alone. Not like it used to be, when he and Sherm and Chub and Jerry and Stan—and Ricky—used to make those runs together.

Link put down the spark plug, lit a cigarette, crossed his legs, and leaned on his knee with his elbow. Those had been the times, all right. Raids on other towns, with all kinds of hell to raise and trouble to get in, and then the race back, which he'd always won. Until the last one. And even then, he'd been first back to Dellville.

Just a year ago. They'd get together every night and go somewhere and do something. Seemed then like the same bunch would stick together forever, having fun. They'd been together all their lives, and they'd always followed his lead, ever since kindergarten. And he'd felt it would always be that way. All of them together, with him showing the way. And now, a year later, Sherm, Chub, Stan, and Jerry were in the army, and Ricky . . . was dead.

And he was left. The time he'd cracked his back in a rollover had kept him out of the army, so he stayed in Dellville and drove Vernham's truck and worked on his car at night.

Link sucked at the cigarette and tossed it away. He was alone, and the guys were gone, and he knew where and why, but it didn't seem right. Any minute he expected to hear them drive up and for them all to go tearing across the countryside. It had always been that way, and he expected it would be that way again, when the others came back.

It didn't occur to Link that the future might be different, or that the boyhood gang had run its course of companionship. He wasn't given to thinking about the future, or about change. His major concern had always been to be Link Aller, and to make the others come to him, and adjust to his leadership. He was satisfied with what he was, and he had no reason to believe he would ever be any different or ever need to be different. When the others came back, his car would still be in front.

Thinking of the raids on Des Moines, and the road races on the way back, he thought of that last race. It was the night Ricky Madison had won the contest in Des Moines for the best street rod. He hadn't really hated Ricky, although he'd had to whip the kid a couple times to keep him in line. People didn't understand that it was Ricky who was always wanting to race with him.

He'd never forget that night. How Ricky had started out first, in that pink-and-copper rod he'd built, with the rest of the gang trailing behind in their cars. He'd been second in line, and he wasn't even trying to race until he realized that Ricky was pouring the coal to his car. Then he'd had to accept the challenge. And although he'd taken after Ricky, he'd never caught up with the pink coupe.

Link uncrossed his legs, lit another cigarette, and relived the race in his mind for the thousandth time. Once before in a race, he had come up behind Ricky and nudged the kid's back bumper. And because he'd done that once, everybody believed he'd done it again the night of the last race, making Ricky speed to get away. But he hadn't. He'd never come near Ricky. Never knew when he came around the last curve on the hill, and across the bridge, that Ricky's car had run off the road, flipped over the end of the bridge, and gone into the river. With Ricky doing better than a hundred when it happened.

Link stared at the ground between his feet, remembering. Everything seemed to have changed after that night. Nobody seemed to understand that Ricky had been his friend, too, and he felt sorry that the kid and his girl got killed. Even the other guys in the gang had quit hanging around with him after that, because they felt he'd pushed Ricky into the race. But it wasn't his fault.

Link spat out a crumb of tobacco, bewildered and angered as he remembered the events that had set everyone against him. He had wanted to show that he was sorry, but from the very first, the town had blamed him for the accident, because he was the leader. And perhaps because he felt deep stirrings of guilt, he reacted to these real or fancied accusations with an air of defiance.

It wasn't the way he had wanted it to be, but he wasn't a boy who knew how to give in or admit a fault. All his life he had been on the defensive, feeling he had to fight for everything he gained. It was his nature to resist every attack with a violent counterattack, and he had done that in reference to Ricky's accident, although it had cost him his friends and earned him the general disapproval of the entire adult population of Dellville.

There had been one opportunity to set things right, but he had missed it. He had gone to Ricky's funeral, keeping a tense, hard look on his face because of the way others were looking at him. At the cemetery he had come face-to-face with Ricky's parents. They had looked at him beseechingly, as though expecting him to say something. He had been in the following car, and they thought he might have something to say. But their grief-stricken staring had been too much for Link. Frightened by the grim finality of the funeral, oppressed by a sense of guilt, on the verge of tears, he had misread the way they looked at him. He saw blame in their eyes and fled. The reason for his abrupt flight was never known to others. All the town saw was that Link Aller had turned his back on Ricky's parents and, in the midst of the final rites, had blasted the hushed ceremony with the raucous thunder of unmuffled pipes as he left.

From that moment on, the town marked him as a vicious, unregenerate boy, and although it tolerated him within its boundaries, it was with disapproval and dislike.

Link knew how he was regarded, but it never entered his mind to leave Dellville. He had been born and raised in the town, he had a job, and as far as he was concerned, he was doing all right. He belonged in Dellville, even if he belonged as its problem. What he felt for the town was something beyond the friendship of other humans, beyond being a member of any group, beyond being accepted by anyone or anything. This was his hometown. No matter what happened on his raids or sallies into other communities, no matter what trouble he ran into in strange places, it was to Dellville that he had always returned at full speed, feeling safe and secure the moment his wheels were on its streets. Nothing really bad had ever happened to him in Dellville, and he felt nothing bad ever would. It was a refuge, albeit a grudging one.

This town was his home, and no matter what frictions existed between him and it, how he defied it, or how it frowned on him, it was his home, and he belonged. It was this feeling, this inner security, this strange level of belonging that made him almost content as he sat alone in his yard and despised everybody.

Link spun the cigarette butt into the darkness. He'd never let on that he had Ricky on his mind. No, sir! He'd stood up to them all and told them where to go. Even, finally, to Ricky's dad and mom. It hadn't been easy to look them in the eye and keep on looking until they looked away first. But he'd done it. He'd looked everybody in the eye. He had to, or they'd see how bad he did feel, and blame him more if he admitted it was his fault. He had to do what he'd done. Tell 'em that you took your chances when you raced, and what happened was your own fault. Nobody'd made him back down.

The only thing was, with everybody gone now, there wasn't anybody to talk to or do things with. He wouldn't mind being shut out of a lot of things if there were other things to take their place. About the only person there was to talk to was old Arnie VanZuuk, and there was a limit to what you could say to that fat old policeman. But he was decent.

The back door of his house opened and slammed shut. He made out his father's figure standing in the light that came through the screen. Going to the icebox on the back porch to get some more beer.

"Link . . . you out there?" his father called hoarsely.

"Yeah, I am."

"You gonna sit out there all night with that light on?"

"I will if I feel like it," Link flared.

"Look at the bugs you're bringin' with that light. The house is full of 'em."

"That's your tough luck, not mine," Link said to the shadow.

"You and that car," his father said, his voice thick and sarcastic.

"You and your beer," Link answered, taking up the spark plug he had put down.

His father slammed the door of the icebox shut and went back into the house, letting the screen door slam behind him. Link snorted. He wasn't going to take any guff off his old man or anybody else. He didn't owe his father anything. He paid his freight at the house. His father . . . seemed like all he could think of when he thought of his father was the sound of that icebox door when the old man went to get more beer. Seemed like he'd been hearing it ever since he was born. Link spat. His father had been half-drunk all his life, it seemed like. He'd been mean, too. Always swinging that belt. Until Link had felt big enough to hit back and had decked the old man. Link laughed quietly as he remembered his father's expression when he hit the floor. They'd had a battle, but he'd whipped the old man. Till he lay on the floor with the sweat and beer pouring off him and he couldn't get up for more. There's been some peace around the house after that. If not peace, at least the old man kept his distance, which was just as good. And as for his mother . . . She just seemed to live with anything that happened, any way it happened. She didn't have much to say to anybody. She did her work and kept her mouth shut and went to church whenever it was open and read her Bible when the church was closed. She was all the time praying the Lord would come and set him and his dad right.

Yeah, if a guy could fight, it didn't matter what anybody thought or said about him. That's what you had to prove to the world. That you couldn't be pushed around. No matter how scared you felt inside, you had to stand up for your rights. Else people pushed you around all the time.

The cars he heard on the street slowed and came to a stop in front of his house. He put down the spark plug and waited to see who it was and what they wanted. He felt his face grow tight and his throat constrict. It was always that way. His first feeling that meeting people meant a fight.

Half a dozen people got out of the cars and walked toward Link. He recognized their voices before he saw them. Kids. What did they want with him? He waited, not even looking their direction as they approached and stood near him. He was so glad to have company he wanted to shout, but he wouldn't let them know it.

"Mind if we look at your engine, Link?" Darrell Atkins asked for the group. A bright-faced sixteen-year-old.

"Go ahead," Link grunted, his head bent over his work. "But watch out you don't mess anything up. Just keep your hands off, you understand?"

"We won't touch anything," Darrell said carefully, not wanting to stir Link to anger. "We just want to look."

Link spat. He hated to admit it, but it was nice having the kids stop by. Even if they were stupid kids, they could talk cars, and he was hungry for somebody he could talk to.

"That sure ain't a regular Chevy engine, is it?" Darrell asked, peering under the lifted hood.

"No," Link said disdainfully. "Any fool can see it's a Jimmy two seventy."

"What else you got on it?" another boy asked.

"Cam, racing pistons . . . well, a lot of stuff you guys wouldn't know if it come up and bit you."

The boys giggled respectfully, and Link felt the old thrill of being the leader again, the way it used to be. Yet, he was suspicious. These kids just didn't "happen" to come around to look at the Jimmy engine. He'd had it a long time, and they'd never come around before to shoot the breeze. What did they want? They wanted something, the way they were shuffling their feet and looking at each other. There was something they wanted off him, but he wasn't going to ask. He was going to sit on his box and not say a word. He was Link Aller, and he didn't need the chatter of a bunch of kids. If they wanted something, they could come to him. Meanwhile, he'd let them see that he didn't give a damn whether they stayed or went. He didn't need them.

Finally, it was the fresh-faced Darrell who, having exhausted his store of car knowledge, came to the point. He squatted down beside Link, looking into the sullen, narrow face. "You heard, Link?"

"Heard what?" Link was aware that the others were very quiet. He was on guard.

"We've got a new cop in Dellville."

"Yeah? What's happened to Arnie?"

"He's still around."

"I don't care if they hire a million cops. I'm not scared of any cops. Never was and never will be."

"This one's tough," Darrell said.

"How do you know?" Inside, Link knew what was coming up, and he felt afraid. But the feeling made him hard on the outside.

"He stopped me a little while ago for the way I turned a corner. And boy, did he give me a chewing out."

"Aw," Link scoffed. "Just because he talks tough, you think he's a hard guy? Anybody can talk tough to a bunch of kids."

"Yeah . . . you ought to see him," Darrell said. "You'd change your tune. Wouldn't he, guys?"

Link stared at them. Kids he didn't want to mix with anyway. Too young. And now they came skulking around because they wanted him to find out how tough the new cop was. Because he never let them forget how tough Link Aller was. A bunch of stupid kids, scared stiff at a bawling out, trying to sucker him into choosing the new cop. Wanting him to lead with his chin.

And yet . . . they had come to him. They needed him. They needed somebody with guts to show them how to act like men, and they had come to him.

They waited, watching his dark, cynical face for some sign.

Link spat. He stood up. His heart was pounding, and his stomach felt queasy, but to others he looked the same—bold, mean, always ready for trouble.

"Some of you guys help me get the plugs back in the engine," Link said, trying to sound casual, "and we'll go find out just how tough Mr. New Cop really is."

When the chips were down, the kids wanted to chicken out. Just the way Link figured they would.

"I don't think we ought to go looking for trouble," Darrell said, looking at the others.

Link closed the hood of his car. "If that's the way you feel, what'd you come here for?"

"To see what you'd found out about the new cop."

"I ain't met him, so I don't know a thing."

"You think he's all talk?"

"Why don't you find out for yourself?" Link asked.

"We don't want to fight him," Darrell said.

"What do you want?"

"Well," Darrell rubbed his shoe in the dirt. "We'd like for you to take a look at him and see if you know him. You've seen a lot of cops. Maybe you could tell us something about him. You know."

"Yeah," Link said sarcastically, "I know. I stick my chin out while you guys hide where it's safe."

"You don't have to. If you're scared or . . ."

"Don't try to give me any of that," Link said. "I know what you're after. I'll put the bell on the cat for you."

"If you think you shouldn't . . ." Darrell's conscience was troubling him. "Maybe it's not right . . ."

"Gotta find out sooner or later," Link said. "Might as well be sooner." Link wiped his hands on a rag and hitched up his pants. "Let's go."

"Where?"

"The schoolyard," Link said, "You guys get off to one side and I'll cut some brodies on the gravel. If he's around, he'll come a-running. Old Lady Taplinger will call him."

"Then what?"

"Then," Link said, "we'll see. Time I get through with him, I'll know all I have to about this tough cop. We'll just see what his line of chatter happens to be."

CHAPTER 5

THE SCHOOLYARD WAS A PERFECT PLACE to play with the car. In the winter it got icy, and it was a good place to come and spin. In summer the gravel was loose, and a good brodie would send up a shower of small stones.

The kids trailed Link to the school and parked along the street, out of the way, where they could watch him. He drove the convertible onto the play area and gunned his engine, getting their attention with the powerful blast from his exhausts.

It was a low-gear game of thunder and screech. He dropped the shift lever into low and floored the gas pedal. The little yellow car shot forward with its rear tires churning back two streams of dirt and stones. Three-quarters of the way across the schoolyard, when his engine was roaring with the sound of a dive bomber, Link lifted his foot from the gas pedal, braked hard as he spun his wheel, then hit the gas again. The convertible slid around, backfiring, roaring, throwing gravel.

He went from one spin into the next. Cutting brodies. Sometimes, instead of braking, he took the slide with the power on all the way through, gunning the engine to make the wheels spin faster and the exhausts roar louder.

Knowing the kids were watching, he was reckless. He sped toward the school building, cut his wheels at the last moment, and slid toward the brick wall until it seemed he had to crash into it. But he had everything timed and planned, and it wasn't new. He made his passes at the building, at the jungle gym, and at the merry-go-round, covering them with dirt and stones and fumes as he drove his car like a bull making rushes at a matador.

He gave his attention so much to his driving that he didn't know the police car had arrived. He didn't notice it until it drove up in the schoolyard, and planted itself in the path of his rush, its lights on him. He had to spin hard to keep from hitting the police car. And when he had skidded to a stop, he cut his engine and waited. This was it.

He sat blinking in the light thrown on him by the police car, trying to see past the light to the man who would get out and come to him, but no one got out. Link scowled, hoping he showed up mean in the light. He felt uncomfortable, knowing the new cop was in the dark, watching him, looking him over, and he couldn't see back.

Well, he wasn't going to sit there all night and be stared at. He shrugged, turned his key and started his engine. He was going to drive away when the siren on the police car gave a brief, quiet growl of warning. Link cut his engine again and waited.

"Come over here, you!"

The voice that came from behind the light was not Arnie's hoarse tone. It was a crisp, hard voice. And tough. Link's stomach quivered.

Taking his own sweet time, Link opened the door of his car, got out, hitched up his pants, and reached for a cigarette. He dallied purposely, to show this new cop he couldn't be rushed. He lit the cigarette and spun the match toward the police car. He went forward slowly, with a swagger, so the kids could see he wasn't scared.

He tried to get a look at the man inside, but when he got up close, he was hit in the eyes by the beam of a powerful flashlight. Link squinted and turned his face away to avoid the glare.

"Don't turn away," the hard voice said. "I want a good look at you."

"It hurts my eyes," Link said boldly. He turned his head away, waiting for the cop to begin bawling him out.

The cop didn't say anything. There was a click, and before Link could set himself, the door of the police car was hurled open, and it smashed against him. It seemed to hit him all over at once, from his head to his knees. He was stunned where it hit against the side of his face and bruised where it hit his chest and legs.

Kern had shoved the door open with his feet. A moment after it slammed into Link with a dull, thudding sound, the door bounced back, and Link had dropped to his hands and knees, his head hanging, trying to get his breath. Kern slid out of the police car with the flashlight on the gasping boy. He reached down and grabbed Link by the front of his shirt and hauled him to his feet. In one motion, he turned Link and stood him up against the side of the police car. Kern's left wrist was under Link's chin, pressing against his throat, Kern's slanting forearm holding the boy pushed tight against the car.

Holding his light inches away from Link's eyes, Kern used his wrist to push Link's chin up and his head back. Link's eyes were glassy. Except for the hold Kern had on him, he would have fallen. His mouth was open, and he was fighting for breath. Kern pressed against him, choking him a little. Link's left eye was beginning to swell and change color.

Kern maintained his pressure as Link sucked air into his throat in long, noisy, tortured gasps. His eyes cleared and his limp body became rigid. He stared into the light that was being directed into his eyes, trying to remember what had happened. He'd said something, and then something had come from out of nowhere . . . He raised his hands and tugged at the arm against his throat.

"Put your hands down!"

Link hesitated. The arm was rough against his throat. He gagged and dropped his hands.

"Don't you puke on me," the cop said harshly. "I'll beat your brains out."

"g . . . g . . . g . . . sick . . ." Link gasped jerkily.

"You're sick, all right. And you'll be a lot sicker before I'm through with you. Won't you?" He shook Link.

"Yeah . . ." Link gasped. He closed his eyes to keep out the blinding light.

"Don't say yeah to me!"

"Yes," Link said weakly, his chin sagging against Kern's wrist.

"Hold up your head! Open your eyes! Yes what?"

Link's head wavered. "Yes . . . sir."

The policeman stepped back, and Link almost fell. He held on to the police car for support. The policeman played the light over him. "What's your name?" the policeman demanded.

"Aller . . . Link Aller . . ."

"Link Aller, what?"

"Link Aller, sir," Link said thickly.

"Don't you forget that." the hard voice in the darkness said. "That's my name to you. Sir. You understand that, you miserable little runt? Sir! Answer me!"

"Yes . . . sir, I understand."

The flashlight beam moved until it was turned on Kern's face. "Take a good look," he said, his face dark and angry.

"From now on I'm running this town, and I want things done my way. I know you," Kern went on, his voice filled with menace as he turned the light back on Link. "VanZuuk pointed you out to me. I know you and your kind. You ain't the first punk I had to straighten out. Do you think I can straighten you out, boy? Do you?"

The light came closer. Link twitched, as though to defend himself. "Yes, sir."

He could see now. The policeman wasn't much bigger than he was, but he looked tough and mean. There wasn't any human look in the eyes. Not even hate. They bored into you, hard and flat. Eyes that could watch you die without blinking.

"All right," the policeman said. "This is just a taste of what you'll git if you make me any trouble. You just git out of line once more, and you won't walk away from it. You know I can do anything I want to you. Anything. You know that, don't you?"

"Yes, sir," Link said. He was feeling better now. Not so shaky, and he was able to be more glib. Better able to bow and scrape while building up hate. "Sir, I didn't think I was doing anything wrong," Link said in a practiced servile whine. "Arnie never cared if we cut a few brodies down here. I didn't know . . ."

A hard hand came out of the darkness and slapped him across the face.

"Don't try any of that on me," the cop said. "I've heard it all before," Kern went on in a jeering tone, "I know your kind. I know why you came down here. You wanted to see what you could get away with. You found out, too, didn't you?"

Link was silent. What had the cop said? Arnie had pointed him out? Arnie. It went to show you couldn't trust any cop, no matter how nice he was to your face. Arnie had put him on the spot. Had been responsible for this.

"Didn't you?"

"Yes, sir," Link said. "I guess I did." His left eye felt as big as a football, and it was beginning to hurt.

"I've had my look at you," the cop said, "and I don't like your looks. And if I don't like somebody's looks, I can be pretty mean if they cross me. You've been a wise punk in this town long enough. I'm lettin' you off easy tonight, cause it's the first time. Next time . . ."

Easy, Link thought. *Letting me off easy*. Cracking my head open with the door, choking me. Easy.

"Git goin'," the cop said. "Before I change my mind and work you over the way you been needin' it."

"Yes, sir," Link said. "Thank you, sir."

He turned toward his own car and limped toward it. He'd only taken a couple of steps when Kern slipped up quietly behind him and gave him a violent kick on the buttocks. The force of the kick sent Link sprawling. He pulled himself into a tight ball, expecting more, trying to protect his head with his arms.

"That was so you wouldn't forget," Kern said, looking down at Link. Kern felt a sense of satisfaction at the boy's instinctive move to cover up. It hadn't taken him long to put the fear of God in this one. The punk would think twice before he stopped long enough to say hello to Darlene. Now that he'd been worked over by the old man, the punk wouldn't be anxious to meet the daughter. Not now.

Kern pointed his light at Link's face, to see the fear and respect. The boy lay on his side, with his knees drawn up, his arms still protecting his face. The side where the door had hit was purple, and the bruise had a split in it that seeped blood. But the boy was watching him with the other eye, and there wasn't any fear in that look. It was a veiled, cold, watchful look, the kind that gleams in the eyes of hurt animals ready for a last effort to fight back if the chance should come.

Kern snapped off his light and walked back to the police car, his feet crunching gravel. He wasn't through with this kid. Not by a long shot. He'd take convincing. The sooner the better.

As Kern drove away, Link got to his feet, brushing off his clothes with trembling hands, trying to breathe through the bubbles of blood in his throat. He looked around uncertainly, moaning with every breath, feeling a desire to cry for help. Something had happened. Something so awful it was almost beyond his comprehension. It wasn't the first time he had ever been beaten, *but it was the first time in Dellville.* His refuge, his sanctuary, was gone. There was no safe place anymore.

The kids drove up to find out what had happened. They hadn't been able to see or hear anything. The sight of them brought forth Link's standard reactions. By the time the kids were within earshot, his moans had changed to short, bitter curses.

"What'd he say?" Darrell asked first as the kids crowded around Link.

"Not much," Link said. He got out his matches, lit one, and held it up beside his face. The kids made sick sounds.

"You got slugged," Darrell said, wincing at the sight of Link's face.

"I been hit before," Link said, spitting.

"Did you have a fight?"

"Didn't have a chance," Link said bitterly, "I went over to talk to him, and before I could say anything, he slugged me with his blackjack. Just sat there waiting for me with it in his hand, I guess. I came up, and wham! He let me have it. Knocked me cold."

"What happened then?" another boy asked.

"I don't know. When I came to, he was kicking me, so I grabbed his legs. I could have upset him, but I was afraid he'd shoot me. I said, 'Quit kickin' me, or I'll dump you on your head.' Then he started swearin' and threatening to shoot me. Even pulled out his gun. But I said, 'Let's not get mad, Officer. If I was doin' wrong, just tell me about it. And put away that horse pistol or somebody'll get hurt.' So, he put away his gun, and he said to quit cutting brodies, and I said if he'd just said that in the first place, that would have been enough, he didn't have to slug me. Then he said he was sorry, and he left."

Somebody sighed with relief. "Then he's all right after all."

"All *right*!" Link protested loudly. "A cop who'll slug you first and talk later . . . all right? Not in my book. His apologizing don't mean a thing. He's a mean snake, who won't give you a chance. But I know him now, and he won't find me so easy next time!"

"We'd better get on home," a boy said nervously. "He might come back."

"What if he does?" Link demanded. "We got a right to talk. Besides," he added, his throat dry, "I ain't finished cutting my brodies."

"You don't dare," Darrell said.

"Don't I? You watch."

"Not me," Darrell said. "I don't want any part of it."

"I only got one more," Link said. "Then I'd better go up to the drugstore and get some ice or something on my eye."

"We'll wait for you around the corner," Darrell said.

"Okay." Link laughed derisively as he started his engine. That cop might kill him if he came back, but Link felt he couldn't quit now. The kids would think he was afraid of the cop. He was, but they weren't going to know it!

Link put his car in gear, zoomed forward, slid his car in the gravel, and headed for the street. He was sweating. But there was no sign of the police car. If it showed up, Link promised himself, that cop would have to catch him to lay a hand on him. And there wasn't a car within a hundred miles that could stay with his convertible. He'd never make the same mistake again of walking to the cop. From now on he'd have to be caught.

The police car didn't show. Link drove into the street, and the boys fell in behind him as he drove slowly toward the town square. He parked and waited until the other boys were out of their cars before he got out of his. Then he got out and walked to the drugstore with his head up, his shoulders swaggering defiantly. Now that it was over, he was proud of his bruises.

Link swung the door back and walked into the drugstore with the kids at his heels. He expected to find Mr. Johnson, the druggist, dozing behind the counter. Instead, there was a strange girl. A blonde, blue-eyed girl wearing a white apron over her dress and looking at him with undisguised curiosity.

CHAPTER 6

Darlene was so bored she felt like screaming. It was half an hour to closing time, and the drugstore was deserted. Even though it was her first day at work, and everything was new, she felt trapped.

She was all alone. Mr. Johnson, the druggist, was in the back somewhere, making up prescriptions, and there didn't even seem to be anybody in the street outside. She turned on the little radio on the shelf near the mirror and tried to get some music, but there was so much static she turned it off again.

A soda or a sundae? She could have anything she wanted. For nothing. Mr. Johnson said she could practice by fixing things for herself. She'd had so much already she'd get sick if she ate any more. Maybe a Coke . . .

She put crushed ice in a glass, let in a generous amount of syrup, and added the carbonated water. She stirred the drink briskly and tasted it. Ugh! It was too sickly sweet. She looked around to make sure Mr. Johnson wasn't watching and poured the rest of the drink down the drain.

A Green River? To take away the taste of the Coke? Why not? Humming to herself, she made herself a Green River and tasted it. She made a face and poured it away in the wake of the Coke.

Maybe she ought to check where the different kinds of ice cream were. She lifted lid after lid, peering inside, taking a bit of each flavor on the end of her finger and tasting it. It was all good and all dull.

She turned to look at herself in the mirror and decided her hair needed attention. She undid the pink ribbon that held her ponytail in place and retied it exactly as it had been before. She looked in the mirror, looked at her creamy skin for any sign of blackheads or wrinkles, tried to see her nose from the side, and bared her white teeth. If that one lower tooth on the side was just a wee bit straighter, she'd have a perfect mouth. Her lips looked too pale. She went to her purse and got out her lipstick, then hesitated. Oh well, why not?

She went behind the cosmetics counter and examined the lipsticks, making a small mark with each one on a piece of white paper, pondering which shades would be good for her. At last, she found a clear, vivid red. She knew she shouldn't use new merchandise, but if she was careful . . . and just this one time . . .

She ran the lipstick lightly across her upper lip, then rubbed the upper against the lower to complete the job. She examined her mouth critically in a small mirror. None on her teeth. And it was a good shade for her. She'd have to buy some when she got paid.

She put away the lipstick and played with the perfume atomizers and bottles of cologne. It was wonderful to be able to try all the different scents without having to buy. She put a few drops of the new scent on her wrist, waved it dry, and smelled. *Mmmmmm.* Exciting. A drop behind each ear. A drop in the little hollow at the base of her neck.

She wondered if there were any boys in Dellville.

There were some boys. She'd already seen two or three. But they were too young. Her own age or younger. And kiddish. Peeking at her when they thought she wasn't looking, and afraid to say hello. Not one with the gumption or the know-how to walk right up and ask her name and for a date.

A date. Where could you go in this dull place? She was momentarily furious again with her father for coming to Dellville. Of all the places to bring anybody to live! And she'd had so many friends in Des Moines, too. Girls and boys who had some life, and had cars and motorcycles, and a gang that was always going somewhere and doing something. Going to drive-in movies and eating places and the stock-car races and doing stuff.

Oh, if she could only be in Des Moines again! There'd be something to do! She could write to her friends, but that was such a dull thing to do. Everything was dull. Oh . . . darn her father anyway, for bringing them to this dead town, just so he could be its policeman. That's all he thought about. What he wanted.

She could always read. She sauntered tiredly to the magazine rack and looked for something interesting. There was so much junk published. So much stuff about politics and war and psychology; it was hard to find a good, interesting love story. Even the confession magazines had doctors writing articles that scared you. Even the comics tried to teach you something. Teach, teach, teach! Didn't a person have a right to read without having to learn something?

Finally, she picked out a movie magazine and, after getting some fresh gum, sat at the counter on the customer's side and began reading an article on happy marriages by a movie star who had had four of them in five years.

Darlene rested her head on her left palm, her elbow on the counter, and tried to read the article, chewing gum, trying to hum between chews, her legs crossed, the toe of the upper foot kicking against the counter. She didn't want to be sitting. She wanted to be out, doing something. Living.

Mr. Johnson came out of the back room and peered at her over his glasses. "You handling everything out here all right? No difficulties?"

"No business," Darlene said. She spoke loudly, because he was a little deaf. "Do you want me to do something, Mr. Johnson? Sweep up or clean?"

"You could sweep up, if you wanted to. You don't have to."

"I'd be glad to, Mr. Johnson. I'll do it now." At least she would be doing something, not just sitting.

"All right, Darlene. If you want to. I'll be in the back if I'm needed."

She swept and pretended that she was dancing, stepping around the broom with her jaws working on the gum, a swing to her hips, a look that she hoped was provocative in her eyes. She looked at the broom from behind her shoulder, giving it the long, limpid look, and joined it in a slow tango, with long, slow steps that moved the dirt and papers on the floor toward the door.

The thought came into her head that someone might see her, and she halted. But what if, instead of somebody local, a movie scout was driving through town, and wanted a soda, and he saw her? Saw how beautiful and graceful and talented she was and signed her to a movie contract? Movie scouts were always touring the country looking for talent. That's what the movie magazines said. And she'd read that most of the stars had been discovered while drinking sodas in a drugstore. That was the first place that movie scouts looked for actresses. In drugstores.

In drugstores in Hollywood, not in Dellville, Iowa. Pooh! She gripped the broom firmly and broke off the dance to sweep with it. And then she heard the straight pipes coming into the square.

She put the broom away, and by the time the yellow convertible had nosed in to the curb at the front of the store, she was behind the counter, in a graceful pose. She could see through the screen door that the driver was alone, and that he seemed young. Then two other cars drove in beside the convertible, and a whole bunch of kids got out. She began to breathe again when she saw they were going to come into the drugstore. At last! At last!

The black-haired boy who drove the convertible came in first. He stalked in with the others at his heels, and her first impression was that somebody she knew from Des Moines must have driven down to see her. He was just like the Des Moines boys, even wearing a black leather jacket on a summer night, western-style denims, and black engineer boots. He even walked like the Des Moines boys—making a noise with his heels, half mad and exciting. And his face even had that take-it-or-leave-it look of a boy who owned his own car. His face . . .

She saw the large purple bruise just behind his left eye, on his temple. A swelling, ugly bruise that had puffed until it cracked and there was a smear of blood around the jagged, blue-black edges of the wound. His left eye was swollen almost shut, his nose bled, and his clothes had dirt on them. Like he'd been in an accident. It was always happening at home in Des Moines, that way . . .

The boy walked along the counter until he was opposite her and then he turned and from three feet away he gave her a long, open, almost insulting look. The other kids hung back a little, but this one was different. He looked at her hair, her eyes, her mouth. She was used to being looked at. She knew how to stand still until it was over. She wasn't going to make the first move. Ask him about his eye, or anything that simple-minded. It was up to him.

And she looked at him, too. Saw that he was older than the other kids—but not too old—and felt his independence, his freedom. He was the kind who did what he wanted. She'd seen others with the same look. And if that was his convertible outside . . . Dellville might not be so dull. He wasn't as cute as he could be. His head was too narrow, and he was awful bony in the face. But he looked lean and strong and . . . interesting. There was a reckless look she liked. If he drove a rod and didn't have a girlfriend already—even if he did—she wouldn't be spending every evening reading movie magazines. So far, in Dellville, he was the best one. He'd be handy to have around until she found better. If there was better. He had a sharp car. And she hated to walk.

He spoke first. "You got any ice I can put on this bruise?" His voice wasn't unpleasant, but it had a hostile ring to it.

"I think so." She didn't move. "Do you want Mr. Johnson to look at your cut?"

"Nah," he said offhandedly. "It ain't much. All I need's a little ice to keep the swelling down."

"A Dixie cup full enough for you?"

"I suppose."

"I'll get it."

She started to fill a paper cup with crushed ice, then hesitated. She looked at the boy. He was watching her hands.

"Wait a minute," she said. She went to the back of the store and got a red rubber bag that was made to hold ice and brought it back to fill.

"Hey . . . look at that," one of the kids who had followed Link in said. "Some service. How do you rate that, Link?"

Link ignored the boy. He waited until Darlene approached with the ice bag. "You're new here, aren't you?" he asked.

"We just moved to town."

"Where from?"

"Des Moines."

"I go there a lot. In that yellow convert of mine. I never saw you there."

"Maybe you didn't look."

"I looked," Link said, grinning.

"You didn't look in the right place."

"What's your name?"

"Darlene. Darlene Kern."

"I'm Link Aller."

She came around the counter to his side. "Lean back against the counter and hold your head back," she said. "I'll put the ice bag on the spot that needs it."

He didn't move, but he looked at her as though he didn't understand.

"Go ahead," she said. "How can I take care of you if you don't cooperate?"

He leaned back slowly, staring at her.

"Put your head back," she said firmly. "You don't have to look at me."

He put his head back, and she had to stand so close to him that her body touched his as she placed the ice bag gently on the swelling. She bent over him, her lips puckered in serious concentration as she tried not to hurt him.

One of the boys groaned. "Boy, if I could get treated like that . . . I'd go out and have the new cop hit me in the eye."

Link jumped. "Ouch! You bumped me with the ice."

"I'm sorry," Darlene said. She wanted to throw the ice bag on the floor and jump on it and cry. It would have to be this way! One boy in town with his own car, and interesting enough to go with, and her father had to beat him up the first night. And knowing her father, she knew what that meant. He would be down on the boy and forbid her to go with him. Oh . . . she wouldn't stand for it. It was time her father learned she had to live her own life. She didn't care what he said. She'd go with this boy if he asked her. If he dared ask, after what had happened . . .

She placed the ice on his eye again, holding it there for him. He looked up at her, and she looked down at him.

Link didn't know what he felt at the moment. He felt foolish, lying with his head back on the counter, with this girl holding the ice on his eye.

But that wasn't it. It was something else, and he didn't know what it was, and in a way he did know. As far back as he could remember, she was the only person who had touched him like this with her hands, to help him. All the other touches had been in fights, or wrestling with squirming girls. But never anything like this from a pretty girl. Taking care of him, touching him to take care of him. As though she cared how he felt.

He stared up at her. "What's your father do?"

She looked at the ice pack. "That."

"Yeah?" Link stared into her face. Yeah! So, this was the cop's daughter. He'd seen her kind before. Nothing bashful about her. The way she looked right back at him, she was asking for it.

The cop's daughter. Nobody else in Dellville would dare make a play for her once Link Aller staked her out. And he was going to do that. Yeah. There was more than one way to get back at a cop for a beating, and this was as good a way as any.

He began cautiously, like an angler making a light cast so he wouldn't spook the fish.

"He don't think much of me," Link said. "I don't imagine he'd care to see you giving me a hand."

"I choose my own friends," she said.

"I guess you're not scared of him, then."

She looked him full in the face. "No, I'm not."

"You wouldn't be scared to have him see you with me? I mean, after I told you he don't like me?"

"If I had a reason to be with you, I don't think I'd care if he saw me or not."

Link changed his position slightly. He hurt all over. It was up to him now. "Do you think I'm scared of your dad?"

"I don't know. It's really none of my business, is it?"

"I ain't."

"Ain't you?" She was still holding the ice bag, but she was waiting to be convinced.

"Want me to prove it?"

"What makes you think I'd be interested?"

"I just wanted to show you I wasn't. Just in case you thought I was."

"A lot of people have been scared of my father," she said.

"Suppose I take you home tonight after you're through here? I guess that would prove I ain . . . I'm not scared of him."

"What makes you think I'd let you take me home?" They were oblivious to the fascinated audience of boys who sat taking in every word.

"Would you?"

"Are you asking?"

"Yeah, if you put it that way."

"The answer is no . . . if you put it that way."

"Suppose I asked you real nice?"

"Suppose you did?"

In this way they began, revenge leading, curiosity responding. Neither was verbally deft, and their opening match was a dogged affair of simple thrust and simple parry. As a conversationalist, Link had the manner and vocabulary of a belligerent, ill-taught boy. Darlene's fund of repartee was a jumbled treasure of standard high school cliches and the dialogue of heroines in second-rate movies. Brought together in the drugstore, the two styles meshed perfectly.

Darlene continued to treat Link's injured face, already making plans as to how she would use him to relieve the tedium of life in Dellville. He represented transportation, and to Darlene, like so many other girls of this automotive age, the wheel was more precious than gold.

Link allowed himself to be cared for and continued his campaign. Knowing from the look in her eyes that it was already won and wanting to laugh exultantly because of the way he had found to make the new cop pay for getting rough. He grinned at the thought.

"What are you laughing at?" Darlene asked, looking at him suspiciously.

"I wasn't laughing," Link said. "I was just thinking of the good times you and me could have if you wasn't too scared of your old man to go out with me."

Darlene's lips tightened. If that's what he thought, she'd show him. "I go out with anybody I want to," she said. "I'm not afraid to go with you, if that's what you think."

"Prove it," Link challenged.

"I don't have to. I know."

"Chicken."

She threw the challenge back at him. "Who'd be chicken if my father saw us?"

"Let me take you home tonight and we'll see."

She looked at him quietly, and he waited for her answer, almost hoping she would turn him down. What if they did run into the cop again? It might be worse the second time. But he'd risk it. He'd be harder to catch the second time.

"All right," she said.

That was how it began.

— — —

After the store closed, they came out with Mr. Johnson, Link having waited inside.

"Warm night," Mr. Johnson said. "Might rain, though."

"Wind's from the south," Link said. "That's where our rain comes from."

"I got my car," Mr. Johnson said. "Can I take you by your house, Darlene?"

"I've got a ride, thanks," she said.

Mr. Johnson looked at Link as though seeing him for the first time. "Yes, of course," he said. "Well, Mrs. Johnson will be waiting. Good night, youngsters."

"'Night," they said back politely.

Mr. Johnson drove away. Darlene couldn't help looking around nervously, but Link forced himself not to look. Not to show any sign of the fear he felt inside.

"That's it," he said, pointing to the convertible.

"Link..."

"You're not going to back out, are you?" Her tone had sounded that way.

"I thought you might put the top up."

"I'm not afraid for him to see us if you're not."

"I'm not afraid," she said. "But why invite trouble?"

"You wouldn't think I was chicken..."

"It was my idea, wasn't it?"

"If it's to please you, I'll do it."

"Thanks. I'll wait in the doorway while you fix it."

So, it was established. The understanding between them was that it was a risk to be together, but it was a risk they both wanted to take. As though they both understood at once that they would be wanting to see each other more and that secrecy and caution was their duty.

"All right," Link called. "Ready to roll."

Darlene made sure the coast was clear, then got into the car as quickly as she could and sank down on the seat to be as far out of sight as possible.

"We don't have to hide," Link said.

"It helps," Darlene said, sounding very mature and knowing. "Let's go before he sees us."

Link backed away from the curb and started forward slowly. "Where do you live?"

"Six seven one Elm."

"Not far from me. I live two blocks north and three blocks west of you."

"Link... do you have to drive around the square, right in front of city hall? He's in there."

"I know it," Link said. "I'm just showing you I ain't afraid. This is the way to your house, and I ain't about to sneak you there some roundabout way because of him."

"You fool, you," Darlene said, but she giggled when they were safely past.

Inside city hall, at the police desk, Virgil Kern heard the straight pipes that he knew belonged to the yellow convertible. He heard them when the car started up across the square, and he heard them as the car came around the square and headed toward the residential section. Kern smiled with bitter satisfaction. That smack on the head had taught that young punk a lesson. He was going around the square at ten or fifteen miles an hour. Probably had lost his taste for skidding around turns for a while. That was the way to keep the punks in line. With an iron fist. Keep 'em scared enough, and they toe the line.

"This is where I live," Darlene said as Link stopped the car. She seemed disappointed.

"I know."

"I suppose I'd better go in."

"Do you have to?"

"I don't have to do anything I don't want to."

"What would you like to do?"

"Move back to Des Moines," Darlene said without thinking.

"Yeah," Link said after a moment's silence. "I guess you would. Can't blame you. It's livelier. And your boyfriend's probably there."

"I didn't go steady," Darlene said. "We just had a bunch that went places together. We didn't pair off like a lot do."

"That sounds like fun," he said. But he didn't mean it. He meant he was glad she didn't have a boyfriend in Des Moines. Or, if she did, she was interested enough in him to deny it.

"Look," Link said. "Do you suppose some night you and me might go somewhere? I'd even take you to Des Moines to visit your friends, if you wanted."

"That's nice of you."

"Would you?"

"We wouldn't have to go to Des Moines. There are other places." Instinct, perhaps, that warned her against trying to mix the two places.

"Sure. Places around to skate, and drive-in movies. How about a movie Thursday night? You free then?"

"I don't know. Could you wait until Thursday afternoon for my answer?"

"Sure. I don't have anything else to do but fool with the car."

"It's a nice one," she said. "It sounds more like a loaded GMC than a modified Chevy. Am I right?"

He stared at her in open-mouthed wonder. "You mean . . . you mean . . . you know about engines and stuff?" He was so delighted he was almost speechless.

"I know a lot of rod benders," she said. "I like hot cars. Any reason why a girl can't? I was going to drive in the stock car races when I was old enough."

"Golly," Link said, overwhelmed. "You are a honey. Imagine finding a girl like you in Dellville!"

"Maybe I can help you work on the car—after Dad gets to know you and likes you," Darlene said.

"I can't see that day coming."

"It might. We could try to arrange it, so he liked you."

"That might take forever. What do we do until then?"

"Oh," Darlene said coolly, "We'll just have to slip around, won't we?"

"You really want to go with me!"

"What's wrong with you? I wouldn't have said it if I didn't mean it."

"Thursday, then?"

"If I can."

"I'll pick you up here."

"Here?"

"Sure." Link grinned. "It's the last place in the world your old man would look for me."

"You're pretty reckless, aren't you?"

"No more than I have to be." He was wondering if he ought to try to kiss her. But she must have been able to read his mind, because she opened the door and got out of the car in one smooth motion, and then said she had to go in.

"I hope you can make it Thursday," Link said, torn between disappointment at not having made a try for the kiss, and gladness that he seemed to have a girl.

"I'm pretty sure I will."

"'Night, Darlene."

"'Night, Link."

She watched him drive away. She wasn't doing so bad. She had the boy with the best car first crack out of the box.

Darlene went into the house. As she stepped through the front doorway, she almost fell over Doyle, who had been crouching there in the darkness.

"Doyle," she said sharply, but in a heavy whisper so her mother wouldn't hear. "You sneak!"

"I'm gonna tell Pa," Doyle hissed.

"Don't you dare!"

"What'll you give me for not telling?"

"I don't have anything."

Doyle's eyes glowed in the faint light. "You do too."

"What?"

"You could give me free sodas at the drugstore."

"I can't."

"I'll tell Pa you're chasing around with the guy in the convertible."

"How about this," Darlene whispered. "One free soda a day, and you can read all the comics free. Maybe I can even get some of the old ones saved for you."

"I can read them anyway."

"Mr. Johnson doesn't like it."

Doyle pondered. "One soda . . . free comic reading . . . and a piece of candy."

"Penny candy."

"All right," Doyle said. "Fair enough. Who's the guy?"

"His name is Link Aller."

"Wow-ee," Doyle said under his breath. "The guy Pa beat up. Will Pa be sore when he finds out . . ."

"Doyle," Darlene said reproachfully. "You wouldn't tell on your sister."

"'Course not," Doyle said. "Not as long as I get the sodas and the candy."

"That you, Dar'?"

It was her mother calling.

"I'm home, Ma."

"About time, too," the tired voice said.

Darlene went in the living room, where her mother was staring sadly at a cardboard box that remained to be unpacked.

"How are you, Ma?"

"How could I be?" her mother replied, feeling contented now that she had someone worth complaining to for a listener.

"Your Pa gone or sleeping, you working, Doyle making more mess than he's worth. All the unpacking's my job. Just like the packing. I don't know why I can't get some help and cooperation from my family. I ain't as strong as a horse. I ain't going to last forever. I'm not strong enough to do all this heavy work by myself. I'm too sick to be movin' and packin' and carryin'. Doctor said I wasn't supposed to lift nothing heavy . . ."

"Then you just sit right down, Ma," Darlene said. "I'll unpack that box. They're my things anyway."

"No, no," her mother groaned. "You've been workin' hard at the store. You go to bed and git your rest, dear. You need it. I'll unpack.

But I don't know how long I can keep up doin' all the work. It's too much for one sick person. Too much..."

She began lifting Darlene's clothes out of the box, sighing painfully with the effort it took to get each garment, trying to explain that she was too sick to work so hard night and day. Breaking off in the middle of a complaint to say, "There's a pretty dress of yours, Dar'. You ought to wear it more. You look so pretty in pale blue. I do remember it seemed a little small last time you wore it, though. I'll make it over for you the first chance I get. If I ever get a chance, loaded down with all the chores and all the responsibilities the way I am, and sick besides. Sometimes I wonder where I get the strength to carry on."

CHAPTER 7

DARLENE WAS WASHING GLASSES behind the soda fountain. Link came in, his eyes searching for her the moment he opened the door. She was alone. He went behind the counter and walked to her side. "Hi, honey," he said softly in her ear. She looked guardedly toward the rear of the store; he checked the front. They were alone. They turned toward each other and kissed quickly, lightly.

Link went to the customer side of the counter and watched her rinse the glasses. "About time for you to go home to supper, ain't it?"

"Five minutes."

"I'll take you."

She washed the same glass a second time. "I don't know, Link . . ." She looked toward the front door. "He might see us."

"What if he does?" There was an angry edge in Link's voice. "He's bound to, sooner or later. We can't hide forever. I don't like the idea of having to sneak around together. I'm not ashamed to be seen with you."

"I'm not ashamed of you, either," Darlene said, resting her hands on her hips, rolling her gum with her tongue. "But he'd be mad."

"He's gonna find out," Link said, "and I'd rather he found out from us. If he don't, he'll think I was afraid of him."

There was real concern in Darlene's blue eyes. "I don't want him to hurt you. He would."

Link's narrow face turned a dark red. He didn't want to have it like that. As though he had to hide because he was afraid of getting hurt. It made him feel inferior to Kern, like he was not a man. It was hard on his pride.

"You said we'd try to break it to him easy," Link said. "If he saw me take you home in the daylight, and that we were friends, that would be the way to start, wouldn't it?"

"Yes," Darlene agreed. "But maybe not right now."

"We've been sneakin' around behind his back for a month," Link grumbled. "That's too long. I don't feel right about it. If he finds out from somebody else, we'd never be able to see each other again."

"I don't know why he doesn't like you," Darlene said, drying her hands. "All I have to do is mention your name, and he gets mad." She didn't elaborate or go into the threats Kern made to her about what he'd do if she dared go out with Link. Link might be scared away, and then she wouldn't have a boyfriend with a car.

"I'm gonna take you home," Link said. "We got to start sometime."

There was a set, defiant look on his face. Feeling anger and fear inside, it made his face look mean. Darlene looked at him and shook her head. Funny how much Link was like her father. Both of them spent half their time brooding about the people who were against them. Both felt they had to go around proving they weren't afraid of anything. As if it mattered. Two of a kind.

If they'd get to be friends, they could sit around and be mad at the world together.

"All right," she said. "If you have to, you can take me."

They went out and got into the yellow convertible. "It would be better if you drove slow," Darlene said. "In case he sees us."

"I have been," Link said. "I've been trying to be a good guy. But he sure watches me like a hawk."

He backed away from the curb and drove slowly in the direction of Darlene's house. They were silent, nervous. Darlene chewed gum. Link stared ahead, his features immobile. It was the first time they'd dared to be seen together. All the other times they had met secretly, after dark, and had driven out of town.

At first it had been exciting and daring. At first Link had felt a sense of revengeful triumph when he'd held Darlene in his arms. He'd done things not so much for the doing, as for the thought of what her father would say if he knew. It was revenge against her father that he felt when he kissed her.

At first it had been a game. Both giggling at the way they'd tricked her father and feeling guilty and excited. Somehow that added spice to everything they did.

And then, one night, Link was lying with his head in her lap, and she was sitting up with a little smile on her face, looking into the night. He stared at the placid peace of her face and the gentle motion of her jaw as she chewed gum with her mouth closed.

And for no reason at all he turned and pressed his face against her chest and held very still. And was aware that she was gently stroking his head. What he felt, and she understood, was beyond play, beyond obvious desire. His need was for something more. Something she had once given him when he had put his head back and she had gently placed the ice bag against his eye.

"Honey baby," she said softly, cradling his head.

Once he would have resented it, but he pressed his face against her, nodding slowly, with conscious effort. And he held her with all his strength. Not to overcome her, but to keep her over him, protective, warm, satisfying in a new way that made him feel humble and like crying.

Then he had kissed her on the lips, gently, for the first time since they had met. And he had driven her home, knowing that it was no longer for revenge but for keeps.

Link felt afraid because he might be getting trapped. Darlene felt afraid because she wasn't sure this was the right one. Even though she liked him. Even though he wasn't good-looking but had something fierce and mean and manly about him. Even though he was always mad at the world just like her father. It wasn't anything she could help or not help. It was just that she couldn't stand boys who fumbled and begged and apologized, but she liked one who was rough and tender, whose touch was firm and not sneaky, who knew how to take over, and knew enough to stop when she really meant stop. And that was Link.

Link coasted to a quiet stop in front of her home. It was a high, narrow house with a steep green roof and was covered with brown asphalt siding marked to look like brick.

Darlene was out of the car the moment it stopped rolling. "Thank you very much for the ride," she said in an impersonal voice. She turned and went into the house, knowing what her father would say if he had seen her, ready with the answers.

Link's immediate impulse was to drive away as fast as he could, while the getting away was good. But he didn't want to act scared or guilty. There was nothing wrong with giving a girl a ride home in broad daylight. He forced himself to light a cigarette and kill a few moments. When he drove away, it would be when *he* felt like going.

He stole a look at the house and saw a movement at one of the windows. He watched, expected to see Darlene, but he met the hostile glare of Kern, who stood at the window shirtless, a cigarette dangling from his lips, a pistol in his hand.

The smile Link had started for Darlene grew rigid on his lips. He stared at the bitter, scarred face in the window, a face that looked back with an expression of barely restrained fury. That look, and the pistol in Kern's hand, were too much for Link.

The gentle start he had planned was discarded by his muscles as they acted on instinct. The clutch pedal came back fast, and his right foot went to the floor on the gas pedal. The little yellow car shot ahead with an explosion-like blast and a screech of tires that left black rubber marks on the street. He took the corner with tires squealing, picking up speed as he turned and straightened, leaving behind the defiant echoes and fumes of his twin exhausts.

– – –

Virgil Kern sat on an old, dark green velour chair in the living room. His shoes and shirt were off, his suspenders crossing over the long cotton underwear he wore both winter and summer. A yellowing home-rolled cigarette dangled from his lips. He was carefully cleaning a stubby .38 caliber pistol.

Mrs. Kern came into the room with an old red scarf tied around her head and a dust mop in her hand. "Never saw a house that got so dusty," she sighed. "If you're workin' tonight, you'd better stop playing with that gun and get some more sleep."

"I slept enough," Kern said. He raised the pistol and looked through the barrel.

"You might give me a hand if you feel so lively."

"And I might not." Kern snapped the barrel in place, lifted the pistol, and brought it down and forward with a slow, steady movement. When he had a bead on the window lock, he squeezed the trigger.

"You already found somebody you want to shoot?" There was a note of mockery in his wife's tired voice.

"Maybe."

"If anybody could . . . Don't you ever think of anything besides shooting and arresting?"

"That's my job."

"You don't have to be a policeman."

"No, I don't," Kern said. "I'd just as soon let somebody else fight the drunks and get shot at by thieves and pick up what's left of people after an accident. But as long as you and the kids want to live in a house and eat three meals a day, I'll go on bein' shot at and fightin' drunks and pickin' up bodies for a hundred and seventy-five a month."

"It's no easy job," Mrs. Kern agreed with a sigh, looking at the lumps and scars on Virgil's face, knowing of others on his body. Knowing of the times he'd come home bloody and bruised, his uniform ripped to shreds. Policemen didn't always win. They were only human and had just so much strength. And when people got a policeman down, they were mean and cruel. Did everything to hurt him bad. Oh, the early mornings he'd come home too nervous to sleep, sick from something he'd seen or from something that had happened. Brooding over what would have happened if that broken beer bottle had hit his throat instead of his cheek. Or if that hatchet had hit him square on the head. Or that knife hadn't been deflected by his badge. People didn't know, who weren't married to policemen. They didn't know what the man took in the way of violence and abuse. Kids teasing him, grown-ups wanting to fight him or cursing him or offering him stingy little bribes. Hurting him every way a man could be hurt and then expecting him to protect them with a smile when they needed his help.

"Maybe it won't be so bad here, Virgil," she said. "It does seem a pretty town. And quiet."

"That's why I picked it," he said. "I already know who the troublemakers are, and I've got them walkin' soft. I had to lay my sap to a couple of heads, but things is quiet, the way I like 'em to be. The way things ought to be. And will be."

"I hope you're not makin' a lot of enemies, Virgil!"

"Can't help it if I am," he said shortly. "My job is to keep order, not win any popularity contests."

She pushed the mop at a speck of dust. "When people get down on you, they get down on us too," she said. "And I did hope I could belong to the local White Shrine and have some pleasure out of it, not have to be hearin' what you did to this one and that one, and bein' snubbed by the ladies because you arrested their husbands or them for some trifle."

"I don't arrest people for trifles," Virgil said, scowling. "They think they're trifles because they did 'em. Runnin' a stop sign ain't a trifle to me, Agnes. I seen too many people killed just that way. There's laws that has to be obeyed, and anybody I catch breakin' the law stays caught. I might not stand around and grin at everybody, but I'm an honest cop, Agnes. Always have been and always will be. Honest. Without fear nor favor. And if our ladies don't like it, they can lump it."

"I'd just like it if I was left out of it," Agnes said.

"Do the same thing back to them," Virgil advised. "You hop on the grocer's wife if he short-changes you or raise hell with the doctor's wife if he can't cure what's ailing one of us. That'll learn 'em."

"It's simpler not to go at all," Mrs. Kern said resignedly.

"I want you to go," Kern said, scratching his chest. "That's what I come here for, Agnes. So you and the kids could have a decent home in a decent town." He frowned, not looking at her. "I want you to belong to all them ladies' clubs," he said sullenly. "Just like anybody else."

"I'd need so many clothes," she said. "I don't have a decent thing to wear as far out of the house as the front porch."

"And the kids, too," he said, ignoring the interruption. "This is the kind of town where kids ought to live. Where it is clean and friendly. Small school like this, they'll get to be in things. Doyle's a husky boy. He'll probably git on the football team. And Darlene's a sure thing to be the beauty queen next year. I've seen the kids in this town. There ain't one can compare with her. Be the most popular girl in school. Have her choice of the nicest friends, too. We can live here. Like human beings."

"I didn't know that," she said. "You never said why you wanted to come here. I thought it was for a better job."

"Worse for me," Kern said. "I might have got some good promotions in Des Moines. But I let that go and come here to take a job because it was good for the family. So you don't have to worry about me makin' a lot of enemies. We're all gonna have a real home here in Dellville. And we won't have to be worryin' about the kids all the time."

"They're good children," Mrs. Kern said, as though she had to convince him of that fact.

"And they'll stay good," he said, spinning the pistol around his forefinger. "We can watch 'em here and see that they don't git into trouble."

"If I had the strength, I'd put on new wallpaper," Agnes said hopelessly. "I couldn't invite anybody to this house the way it looks. And if you join a club, they expect to meet in your house too, sometime."

"I ain't ashamed of it," Virgil said. "If they want us to live high, they can raise my pay."

He looked at his wife as she stood leaning against the wall. She'd been a pretty enough girl, but she'd changed. Maybe it would help her to join one of those silly ladies' clubs. Give her something else to think about besides how bad she felt. If she'd put on something decent, and stand up, and do something with her hair, and fix her face, she might not be a bad-looking woman now.

Suddenly he got up and went to his wife, taking her hand. "I know it ain't been easy for you," he said. "Bein' a policeman's wife, and me always workin' nights. All the work and the care of the kids to worry about. I know you're a hard worker, Agnes. Maybe things will git better here, and you won't have to work so hard."

"It's all right," she said. "I don't mind working hard if I think it's appreciated once in a while." She raised his hand to her cheek. "I know your job is awful hard, Virgil. You're a good man."

He bent down and kissed the top of her head, feeling the coarse, straight strands against his lips.

"You're pretty sweet," he said in a low tone.

She broke away. "Oh, dear," she said. "There's a car stopping outside."

"Let 'em stop," he said. "I ain't expecting anybody."

"It's Darlene."

He looked toward the window and caught a flash of yellow. For a moment his head felt as though it would burst. "It's that Aller kid," he said, breathing hard. "I'll learn that pup to hang around her!"

"Don't get upset, Virgil. The boy just brought her home."

"I know him," Kern repeated. "I know him. I knew he'd find her. I'll kill him if he lays a hand on her . . ."

Darlene ran in, her eyes looking from one parent to another. She looked guilty, evasive. Her eyes were too bright, her face flushed. It was because she was scared, but they didn't know that. All they saw was an excited girl who looked guilty about something.

"Hello Ma . . . Pa . . ." she said. "I'll be right down to help as soon as I change. I . . . uh . . . Link Aller was nice enough to bring me home. He came in the store just as I was leaving. Be right down."

Kern picked up the empty pistol and walked to the window. He stared at Link. The black-haired boy stared back, grinning insolently. Then he gunned his Chevy and took off in a cowboy start, leaving a thin trail of blue smoke hanging in the air like a visible puff of contempt.

For a moment Kern felt a terrible sense of despair. After all his planning, all he'd given up to protect his family, he'd delivered his daughter right in the arms of the worst punk in town.

No . . . no! Kern brought up the empty pistol, drew a bead on the fleeing convertible and squeezed off a dry shot. No!

CHAPTER 8

Link didn't think anything about it when the '49 Lincoln passed him on the street and cut in front of him. He was making deliveries in the Vernham's Market panel truck, and his mind was on his work. He wanted to get it done so he wouldn't be late for his date with Darlene.

He had to jam on his brakes to keep from hitting the Lincoln. It had come to a stop right in front of him, blocking him. He was going to yell something at the crazy driver when the driver jumped out of the car, and he recognized Virgil Kern wearing civilian clothes. Link felt his stomach muscles grow tense.

Kern came back slowly, purposefully, his black hat low over his eyes. He stood at Link's window and stared in. "Lemme see your driver's license."

"I didn't do anything wrong," Link said.

Kern's head came up an inch. Link didn't know what he was expected to do. "Sir," he added, reaching for his license. Kern took the license and read it as though he had never seen one before. Link waited him out tensely, swearing to himself because of the time he was losing. But Kern took his time studying every word, reading the violations on the back.

"You been in a lot of trouble," Kern said.

"No more than my share . . . sir." How he hated to say that.

"The law don't give anybody a share," Kern snapped. "You're inside it or outside it."

"Yes, sir," Link said.

"You ran a stop sign," Kern said.

"No, I didn't. I came to a full stop. I know I did."

Kern stared into Link's eyes. "I saw you do it."

"Honest," Link said, "it wasn't me. Not this time."

"You'd better come down to the station with me," Kern said.

Link shook his head helplessly. "I've got a lot of deliveries to make. Can't I come some other time . . . sir?"

It was like begging a cobra not to strike. "You comin', or do I have to bring you in my way?"

"I'm coming," Link said dully. "But I know I didn't run that stop."

"I might as well tell you something," Kern said. "It's your word agin mine. We git down there, you'd better pay your fine and keep your mouth shut. I wouldn't want you to call me the same thing as a liar in front of the mayor. It might make me mad."

Link didn't answer. There was no answer to this. He followed Kern to city hall, to plead guilty and pay his fine. And he wasn't guilty. He wasn't.

"Ach, Link, you should know better than to run stop signs," Arnie VanZuuk said when Link pled guilty. Arnie was in uniform, sitting at his desk, watching.

Link was tempted to tell the fat old man the truth, but he remembered Kern's warning and remembered that it was Arnie who had pointed him out to Kern. He kept quiet. You couldn't say anything to any cop.

"You better watch yourself," Kern said as Link got back in his truck. "I'm gonna be watching you, boy. You ain't going to get away with a thing from now on." He added significantly, "Not in this town, Aller."

Kern watched Link drive away. It was the first time in his life he had ever pulled anybody in on a phony charge. It made him feel dirty. Sneaky. It made him hate Link all the more for making him do it.

Virgil drove home feeling guilty and angry. But a man had to protect his daughter. And Dellville would be better off without Link Aller in it anyway. A record as long as your arm. Reckless driving, fighting, smuggling cigarettes and firecrackers in from Missouri. Just a small-town punk. And Darlene had to pick him.

— — —

There was a look of cold displeasure on Rudy Vernham's meaty red face when Link drove up to the back of the store in the truck. That boy! He was getting worse every day. Bad enough when he was always driving the truck like a hot rod, but at least then he got his work done. Now he was always late, always behind. Customers were calling and complaining. Where is Link with my order? I can't start dinner!

Where was Link? Who could tell? In some trouble, probably. Racing with somebody in the truck, even. Until some day he would kill somebody, and the store would be sued.

And you couldn't talk to him. He did his work all right, but he was sullen. Look at him now, with that mad look on his face. That would make friends for the store, all right. Sure. A lot the boy cared if he drove away business. That was the trouble with the younger generation. All they wanted was thrills. They didn't care about taking real responsibility for a job. The store would be better off with an older man driving the truck, who knew how to smile and handle complaints. If he could find one who would work for Link's wages. Problems . . . problems . . . people didn't know . . .

"Don't sit down," Rudy called as Link entered the store. "I've got a truckload of orders waiting."

"I'll get 'em," Link said sullenly.

"What takes you so long?" Vernham asked. "You should have been back an hour ago."

"Same old thing," Link said. "Kern ran me in."

"You should be more careful. That's a truck, not a race car."

"I am careful Mr. Vernham," Link said, a harassed look on his dark face. "Honest, I am. But that cop is on my tail all the time. I bet he stopped me four times today. And every time he stops me, he's got to read my license, and stall and stall. He knows he's making me late. I don't know. For some reason or other he's always picking on me."

"Yeah, sure," Vernham said. "Picking on you. That's what you all say. Somebody arrests you for speeding, and he's picking on you."

"He is when I'm not speeding," Link said. "He makes up those charges. They aren't true."

"Plead not guilty, then."

"I can't," Link said. "He'd beat me up."

Vernham shook his head. What imagination these kids had! "Look, Link," he said. "Don't try to kid me. I know you. I know how you drive. You've been pretty careful with the truck, but I've seen you moving pretty fast around town in it."

"You have to move fast to make all those deliveries," Link said.

"Sure. And sometimes a cop sees you, eh? I know, my boy. I know. You just have to keep a sharper eye out for the cops."

"Or drive around in low," Link said.

"We can't be late with the orders, Link. Now, load up and get these orders out. The people will be getting impatient. Don't waste time, eh, Link? Stay away from the drugstore. You can see her after working hours."

"Her?"

Vernham laughed. "It's a small town, Link. Everybody knows."

"Her father, too, I'll bet," Link said. "That's why he's after me."

"As good an excuse as any," Vernham said jovially. "Now come on, hot rod, let's get those groceries delivered."

"I will if Kern lets me."

"Don't blame somebody else for your troubles," Vernham said sharply. "I know Mr. Kern, too. Better than you. He's a very nice man. A little strict, but very nice. He has the town's welfare in his heart. I know that from talking to him. You do your work, and pay attention to your driving, and you'll have no trouble. I know that. I know it."

You couldn't win, Link thought with weary anger as he loaded the truck. *They were all against you. Just because you liked to mess with engines and drove a rod. And had been nailed a couple of times for speeding and stuck with a reckless driving charge once or twice. Mainly because you had a yellow rag top with straight pipes. What the hell did they want you to do? Curl up and die?*

He looked at his watch. He'd have to go to get all the orders delivered. If he didn't, they'd be calling Rudy about it, and he'd catch it. If he tried, Kern would be on him. It had been that way for weeks.

Link got in the cab of the truck and backed into the alley. He drove out on the street in low, shifted into second and put his foot on the gas pedal. The truck roared away, the engine giving a quick scream as he slam-shifted into high without taking his foot off the gas. He smiled grimly. It wasn't everybody who could slam-shift a truck and get away with it.

In one drive and out the other, dropping his boxes and packages, not waiting to hear complaints, his mind filled with the wrongs he had suffered at Kern's hands and his plans for revenge. If only he wasn't Darlene's father! Even if he was, he'd get what was coming to him if he didn't back off. Darlene wouldn't care. She didn't like the way her father acted either. Treating her like dirt too. Just a chance, Link thought, making a fist. Just one good chance! Teach that cop a lesson, and he wouldn't be so anxious to pick on a guy. But a guy had to fight back if he wanted people to leave him alone. He couldn't wish for it. He had to fight to be left alone.

A wry grin parted Link's stiff lips. He'd had to whip his own father to get some peace and respect around his own home. Maybe he'd just have to whip the man who was going to be his father-in-law too, for the same reason.

In one driveway and out. And another, and another. As fast as he dared to go, with an eye open for Kern all the time. He was in a hurry, afraid to hurry. If Kern was tailing him, he couldn't give the cop a real excuse to stop him. Link knew he'd have to make his stand when the cop was in the wrong, throwing his weight around. That's when he'd have to whip the cop and show the kids in town who Link Aller was. He'd cut that cop down to size and marry Darlene and stay right in Dellville and drive the truck the way he wanted to and be somebody. Somebody who didn't need anybody else, somebody who got along fine and was let alone.

If he hadn't been so busy looking for Kern, he would have seen the little girl. He wasn't going very fast, but he was looking toward the corner, and not in front of the truck. And when the little girl ran out from between parked cars, he was a second late hitting the brake. He felt the bumper hit the girl at the same moment he had the truck stopped.

Link was out of the truck like a flash, darting around to the front. The little girl was lying on the pavement, under the front bumper, crying at the top of her lungs.

"You hurt, kid?" Link asked anxiously, on his knees. He reached for her and got her out in the open. She didn't seem to be hurt. He had only just bumped her a little.

A woman ran toward them, screaming.

"I think she's all right, lady," Link said. "She ran out in front of my truck, and she got bumped."

"Oh my God! My little girl's been run over. Call a doctor, someone. Hurry!"

A score of people had gathered by this time. The word leaped back and forth. Little girl. Run over by truck. Link Aller's truck. Aller? No wonder. The way that boy drove. Bound to kill someone sooner or later. Look what he did to poor Ricky Madison and Sharon Bruce. And now a little girl. Awful. No surprise, though.

Link stood at bay by the front of his truck, his heart heavy. This was it. This had to be it. What Kern was waiting for. Arnie drove up in the police car; he called Link to him and asked for his story. He listened, nodding. "These things happen," he said slowly. "With anyone else, it's nothing. Everyone is glad it was no worse. But it's bad for you, Link. Bad."

"Yeah," Link said. "I'd better finish making my deliveries while I can."

"Better drive the truck to the store," Arnie said. "Better you shouldn't make any more deliveries today. You don't look so good."

"I don't feel so good," Link said. "If I'd been going a little faster . . ." He shuddered.

"Remember that," Arnie said. "The next time you want to drive fast. Remember."

Link drove the panel truck back to the store. It was lucky for him Arnie had shown up. If it had been Kern . . .

When he got back to the store, Vernham was waiting for him, his round red face twitching with anger. "So that cop was picking on you!" Vernham bawled. "He should have picked harder. Those parents are going to sue for plenty, and I'll have to pay."

"It couldn't be helped," Link said. "She ran right in front of me."

"Sure," Vernham said sarcastically. "I know. She was picking on you too. I should have known better than to hire you in the first place. You're fired. Right now, this minute. Fired!"

CHAPTER 9

It was almost closing time in the drug store, but Darlene wasn't alone when Link came in. Her brother Doyle was sitting on the floor by the magazine rack, reading a comic and drinking his free Coke.

Link went to the far end of the counter and sat down. He was wearing a white shirt and dark trousers. His hair had been combed, and it gleamed wetly. He had shaved. Darlene looked at him approvingly. Link wasn't a bad-looking boy when he cleaned up. She was wearing a sheer pink nylon blouse over a sturdy white bra. She knew Link thought it was too daring, but a lot of the girls in Des Moines dressed that way, and she didn't want to look like a country cousin. She had on a white-and-gold cotton brocade skirt, pink anklets, and white summer shoes.

Link leaned both elbows on the counter and sighed as though for the first time that day he could breathe out.

"I guess you heard what happened today," he said dispiritedly. Darlene nodded. "I know it wasn't your fault, Link. Was it?"

"She ran right in front of me. I didn't even see her."

"You poor dear. Want a Coke or something?" She wanted to do something nice for him. Whatever she could. He nodded.

"I got fired, too."

"You'll get another job," she said in a coaxing, reassuring voice. She gave him a Coke with a lot of syrup in it.

"I don't know where," he said moodily, playing with the glass. "I tried some other places. Nobody wants to give me a job that's got any driving connected with it."

"Take another job."

"I don't want any other kind," he said stubbornly. "I'm a driver. That's my trade. And it wasn't my fault, what happened. But try to convince anybody. Trouble is, everybody in this town is down on me. I've got a good mind to pull up stakes and go somewhere else."

"What about me?" she asked, her eyes wide.

"You could go with me—if we was married."

"Not so loud," she warned. "That's my little brother."

"I know. He ain't listening."

"You don't know Doyle."

"What do you say? Are you game?"

She moved closer to him, leaning against her side of the counter. "I don't know." She looked troubled.

"We've talked about it. I thought you said you wanted to."

"I do. I do. But I didn't think so soon. I'm only sixteen."

"You'll be seventeen in a week or so. That's old enough."

"Gee . . ." She twisted a straw in her fingers. "I don't know, Link. I want to get married, but I want to have some fun first."

"Who with?" he demanded.

"With you, of course, silly," she said in a low voice.

"We could have more fun if we was married."

"I'm not through school yet . . . I'll miss all the fun at school.

Pa says I ought to be picked for the beauty queen next year. Wouldn't you like to wait, and be married to a real beauty contest winner? I've never won anything like that in my life."

"Well," Link said stiffly, pushing the Coke away. "If that means more to you than I do, all right."

"That's not it at all, Link."

"It's the way I see it. Guess I'll have to leave town alone, then."

"Link . . . you could stay if you'd take another kind of job."

"I might," he said grudgingly. "If I was sure you was gonna marry me."

"We'll talk about it later," she whispered. She reached out and furtively touched his hand, ready to spring back if Doyle raised his eyes from the comic book. "Okay?"

"Okay." He reached for the Coke again, his eyes sad.

Darlene walked toward her little brother. "Time for you to run home, Doyle."

He answered without looking up. "I thought I'd wait and walk home with you, Darlene."

"I can walk home alone all by myself."

"Yeah . . . but I don't like to go alone. I get scared."

"Doyle Kern! What a lie!"

"I do. Honest." He grinned mischievously.

Darlene looked helplessly at Link. She didn't dare threaten Doyle or put him out. He'd tell on her. Link strolled toward Doyle, hands in his pockets. Doyle settled himself for the bargaining.

"I was just thinking, Doyle," Link said. "You look like a kid that would have fun with firecrackers."

Doyle forgot his blasé pose. "Have you got some real firecrackers, Link?"

"I know where to get some," Link answered.

"Huh. So do I. In Missouri. What good's that?"

"Missouri ain't so far," Link said. "I'd take a little run down to Missouri any time for a friend."

"For me?"

"If I thought you was a friend."

"I'm your friend, Link. Ain't I, Darlene?"

"Tell you what," Link said. "You go home by yourself, and I'll go to Missouri tonight and buy you a great big sack of firecrackers. What do you say?"

"A real big sack full?"

"Real big."

Doyle got to his feet. "Don't you forget, now. Tonight."

"Tonight," Link said.

"Is Darlene going with you?"

"Why'd you ask that?"

"It'd be more fun if you'd take me, and I could pick the firecrackers I want."

"I'd get you home too late," Link said.

"That's all right. Dad don't care what time I come home."

"I'll get you some of every kind," Link promised. "If you just run home and wait."

"All right," Doyle said. "I'll do it if I can have a nickel candy bar now." His calculating glance went from Link to Darlene. She shrugged and went to the candy counter. Doyle took the candy bar from her and sauntered out, the comic book under his arm. He knew Darlene wouldn't call him back to return it.

"The little racketeer," Link growled.

"You're the one who made all the big promises," Darlene said, scolding slightly.

"I had to get rid of him. I want to be with you tonight."

"Are you really going to Missouri?"

"Sure. It's only a forty-minute ride. We'll get the fireworks and come back. We can talk while we're riding."

"All right."

"Leave from here?"

Darlene shook her head. "I'm afraid Pa might see us. I'll walk to the edge of town, like always. And you'd better leave now. Pa wouldn't like it if he found us here alone."

"It's a store. He can't make me stay out of a store."

"Be stubborn, then," Darlene said. "But if he comes in, he'll be mad, and he'll take me home."

"Okay," Link grumbled. "The edge of town." He went out feeling like he had been shamed and humiliated in front of Darlene—that he had been proven inferior by fleeing the name of his enemy, not even the visible threat.

- - -

They drove back from Missouri with the paper sack of fireworks on the rear seat. Darlene sat close to Link, with her head on his shoulder, her arm around his waist. He drove with one hand, right arm around her shoulders, holding her close. He was leaning against the car door with his shoulder, and she half-lay on him. It was comfortable, and he was in no hurry to get back. He drove at sixty-five or seventy, dreamily listening to the mellow tones of the mufflers.

They had talked it all over on the way down, and it was settled. Link would look for another job in Dellville, and Darlene would finish school. They'd go steady, as they had been doing, and hope her father would get used to the idea. Even if he didn't, they'd get married as soon as Darlene was out of school—or sick of it—and they'd go somewhere else, where Link could get another job driving a truck or something. Nobody could do anything to stop them once they were married.

Link was pleased with the arrangement because it would give him time to save a little money and buy some of the speed equipment he wanted before he had to save for being married. He still resented the idea of being made to take a non-driving job, but maybe, after a couple of months, he could get another one. Driving was his trade, and nobody had a right to keep him from it.

Darlene liked the arrangement too. It would give her time to finish school, maybe be the beauty queen, and get to know some of the other kids better. She thought Link was the leader of the boys at first, and that's why she'd picked him. Now she had found out that he didn't belong to any crowd at all, and it wasn't too much fun just going alone with him all the time.

Of course she liked him, and if she had to stay in Dellville, she'd probably marry him. But it was more fun to tear around with a crowd and go places and have fun than it was to be alone, just the two of them. It was better than anything else in Dellville so far, but some of the other boys seemed cute. The only way she'd ever be able to know any of them would be in school. They were afraid to be friendly with her outside because of Link. Darrell Atkins had told her Link had threatened to beat up anybody who hung around her. That wasn't right. It showed he loved her, she guessed, but it wasn't much fun.

When it came to fun, the other kids seemed to be having more than she and Link had. But she could promise to marry him, and maybe she would, and meanwhile she could finish school and win the beauty contest and have some fun.

"Want to stop somewhere for a couple of minutes?" Link asked.

"What time is it?"

"Not very late. Not even eleven."

"We'd better not. It's too late."

"Just for a minute . . ."

"It's never a minute. We'd better not, Link. Not tonight."

He sounded hurt. "You'd think we could stop for a minute on the night we decided to get married."

"I want to stop," she said. "But I don't dare be too late. You won't want to leave when it's time for me to go home."

"Cross my heart I will. Whenever you say so."

"I say now."

"I mean whenever you say so—after a while."

"Promise?"

"Promise."

"All right, then," she said reluctantly. "But just for a minute."

He was silent as he scanned the countryside for signs of a side road. His lights picked up a sign warning of a junction. He slowed and made the turn onto a gravel road. A mile or so along, he swung off on a narrow dirt road and turned off it to go through a gate that led into a wooded pasture. He cut the engine and turned off the lights.

"How did you know about this place?" Darlene asked.

Link chuckled, pleased at the note of jealousy in her voice. "I found it while getting away from the Highway Patrol one time. Like it?"

"It's so quiet," she said, leaning toward him.

Link leaned back and put his arm around Darlene. "All we need is five hundred dollars," he said. "We get that saved, and we can get so far from here . . . Just enough to live on until I find me a good job wherever we go. California, New York, or anywhere."

"I've never been out of Iowa," Darlene said. "I wonder what it looks like in those other places."

"You've been to Missouri," Link teased.

"Missouri! That's just like Iowa, only they sell fireworks."

"I'll take you anywhere you want to go," Link said. "You name it, and we'll go there."

"I'll think real hard, and you think real bard," Darlene said. "Let's see if we both think of the same place."

"Okay."

She rested her head on his shoulder, and they searched their minds for the name of a promising land. She remembered places she had seen in the movies, where there was sunshine and beaches, and everybody wore nice clothes and lived in nice houses, and she wondered what the names of those places could be, and where they could be found. Link thought of wide roads, speed shops, and saw himself in coveralls working happily on engines in . . .

He woke up first and knew by the angle of the moon that it was late. He was cold, and his arm, still around Darlene, ached miserably. But the moment he realized they'd been asleep, and it was very late, he was scared. He shook Darlene, trying to wake her. Her face looked swollen, and in sleep her lips protruded in a babyish pout. She didn't want to be awakened and tried to burrow against his shoulder. He kept shaking her. She came to slowly, and he saw the sudden fear that came into her eyes. She sat up, wild-eyed. "Oh my God," she said. "What time is it?"

Link looked at his watch by the light of the dash. "It's after two."

"Oh my God," she repeated, her voice hoarse with apprehension. "Oh my God."

"Let's get out of here," Link said.

He switched on his headlights and turned around on the rough ground and headed back to the main highway. He drove with both hands, as fast as he dared. When they hit pavement again, he floored the car, pushing it home at a hundred miles an hour. Darlene sat bent over, hugging herself with her arms, shaking with cold and fear. She didn't speak until he had to slow down for Dellville.

"Let me out on the edge of town, Link. He'll hear us."

"I'll go the long way around," Link said. "Won't go through town at all."

He went slowly, his eyes alert, for the first time in his life cursing the throaty roar his duals made at low speed. But the streets were deserted, and in a few minutes he coasted to a stop in front of Darlene's house.

"Made it," Link said. He was anxious for her to get out, so he could get away. He tried to make their good-night kiss a quick one, but she put her arms around him and held him close.

Suddenly her face was no longer in the dark but illuminated by the powerful beam of a flashlight.

– – –

At midnight, Virgil Kern stretched, yawned, and decided to make his rounds. He climbed into the black police cruiser and idled through the dark streets of town. Everything was quiet, deserted, peaceful, as it should be. There were lights on in a few houses. Sick babies, he guessed, or hungry ones. But nothing moved on the streets. The way it should be. If there was anything wrong going on in Dellville, it was going on quietly. So, it couldn't be too wrong.

This was the town, all right. No fights, no drunks, no brawls. No getting scared every time he saw a parked car. The kids in this town went to bed early. Just to reassure himself, he drove close to a couple of cars and flashed his light inside. They were empty. As they should be.

He drove down the street where he lived. His house was dark too. Agnes and the kids would be asleep now. He had a mental picture of how they would look. Agnes on her back, stiff as a corpse, her thin nose pointed at the ceiling. Darlene in pajamas, looking soft as a puppy as she slept with one arm thrown over her old teddy bear. And Doyle, whose bed would be messed, covers all twisted and half on the floor, head and feet sticking out, a funny puckery look on his face. That Doyle. He'd look like a rat making a nest in a pile of papers. Kern chuckled.

It wasn't so nice to be away from the family six nights out of seven. A man missed a lot that way. But it was good to know you were patrolling the streets to protect them. Good to know that all the people depended on you for protection. It made you feel like a man, to know that people were sleeping easy in their beds because they knew you were on guard. All those solid businessmen, and kids, and old maids, and young women, all curled up soft and comfortable in their beds.

He frowned as he drew up even with his house. There was somebody sitting on the front porch. He stopped the car and turned the big light that said POLICE across the glass to focus on his porch. Doyle's face blinked and grimaced back at him.

Kern turned off the light and got out of the car. He'd stop and chat with Doyle for a minute. Get to know his son a little better. Didn't see the boy much. Not much chance to talk and be pals and find out why the hell he wasn't in bed at midnight.

"Doyle, what're you doin' up this hour?"

"Nothin', Pa."

Kern sat down beside his son. "Nothin', eh? You'd better be doin' nothin' in your bed. Your ma catches you up like this she'll whale the tar outa you. Can't you sleep?"

"I guess I could," Doyle said, yawning. "But I gotta wait."

"For what?"

"For somethin', Pa. It's a secret."

Kern glanced up at his house. "How long has your ma and Darlene been asleep?"

"I don't know," Doyle said. "I went to bed before they did."

"What in the world are you waitin' for out here? It's a mite late to see the Easter rabbit, and a mite early to see ol' Santy Claus. You better run up to bed now, Doyle."

"I can't, Pa. I gotta wait. I promised."

"Who promised what, son?"

"I can't tell you," Doyle said. "You'll whip me."

"Maybe I will and maybe I won't. But I will if you don't get up to bed or tell me what this is all about."

"Would you be mad if I had some firecrackers, Pa?"

"They're against the law in this state."

"See," Doyle said. "I knew you'd be mad. And now I won't get to have any firecrackers after Link went all the way to Missouri to get 'em for me."

"What'd you say, boy?"

"Nothin', Pa."

Kern gripped Doyle's wrist. "You'd better talk to your pa, boy. Why is Link Aller bringin' you firecrackers?"

"Fer a present," Doyle whined.

"What fer, fer a present? Talk, boy." Kern increased the pressure on Doyle's wrist. Doyle yelled aloud. "I ain't hurtin' you, but I will," Kern said. "What have you got to do with Link Aller?"

"Nothin'," Doyle wailed. "He just said he was gonna bring me a sack of fireworks, and I'm waitin' for 'em."

Kern looked at Doyle for a moment, then dropped his hand. He got up and went in the house, lighting his way with the hand flash. He went up the stairs quietly and looked in his own room. Agnes was asleep on her back. She was alone. Kern went to Darlene's room, hesitated in front of her door, then opened it, flashing his light on her bed. It was empty. Still made up.

He walked downstairs again and went out on the front porch. "Go to bed, Doyle," he said.

"But Pa! My firecrackers!"

"You go to bed," Kern said. "I'll see Aller when he comes back, and I'll git your firecrackers. I'll git your firecrackers . . . !"

He went back to the police car, got in, and started the engine. And he began to hunt for them.

He drove a while, then stopped a while to listen, then looked some more, then stopped again and listened. He was hunting, and the hunt was professional at this stage, impersonal. It was his job. He almost enjoyed it.

Shortly after two in the morning, he heard them coming into town the way he knew they would. The long way around. He drove back to his street and ran the police car into the driveway across the street from his house. He waited behind a tree as the yellow convertible made a stealthy turn at the end of the street and coasted quietly to a stop before the Kern house. He gave them a minute, and then he started forward noiselessly. He reached the side of the yellow car and pressed the switch of his flashlight.

For a moment no one moved. Kern held his light steady, Darlene stared into the beam, and Link was rigid with terror at having his back to his enemy. Then Darlene said, "Pa," in a tone of defeat and despair. And that broke the spell.

Link tried to climb over Darlene to get out the far side of the car. But even as he raised up, Kern grabbed him by the hair and an ear and pulled him back. Darlene screamed.

Link fought naturally, like a cat. He didn't have to think or plan. He'd been in too many fights to need that. He didn't resist the pull but gathered his feet under him and shoved with all his might. His head rammed into Kern's chest, and Link heard Kern grunt as the wind was partially knocked out of him. He let go of Link's hair and ear, and Link fell across the car door, his body slanting toward the pavement, trying to kick himself free with his feet.

But he hadn't hit Kern hard enough to stop him for more than a moment. Link was still hanging head down when Kern rushed him. He fell to the street, curling himself up in a tight ball as Kern tried to drive his foot into Link's groin. Link smothered the kick, taking it on his arms as he grabbed Kern's foot and tried to upend the policeman. But Kern was expecting that, and he kept his balance as Link lifted his leg. Kern hopped on the opposite foot, trying to smash at Link with his blackjack while Link dodged and tried to trip the policeman.

Darlene had come out of the car, and was watching, her hands pressing the side of her head, her screams blending in one loud shriek of terror. The two men fought quietly, grunting when they struck or kicked, but saying nothing. It was too fierce for words. Too late.

Kern pulled back his right arm and felt someone grab his wrist. He turned and saw Darlene clinging to it, wild-eyed, still trying to lift hoarse, rasping screams from her throat.

Kern let go of Link and seized Darlene. She stumbled to the front of the car, trying to scream for her mother. Lights were on in every house along the street. People were running out in their night clothes, fear and curiosity in their faces.

"Inside!" Kern snarled hoarsely at Darlene as he heard people running toward them. "Inside!"

He turned back to Link, blackjack raised. He would beat and beat and beat . . .

He lunged forward to deliver the blow, but it never landed. Arnie VanZuuk, routed out of bed by frantic telephone calls, had finally arrived. The big, fat old Dutch policeman wasted no time or breath. He bore down on Kern like a tank, and hit him a battering-ram blow with the big, hard belly. Kern flew through the air and fell stunned to the street. Arnie followed him and took away his gun and blackjack.

Then he bent over Link, shaking his head, muttering to himself.

"Somebody, there!" Arnie bellowed to the crowd. "Send quick for the doctor. Quick, quick. You others go home. All is over now. All is over."

There was no hospital in Dellville, no ambulance. Doc Frommer took Link back to his office in his own car, and Arnie took Kern there in the police cruiser.

For a while, some of the people stood around and looked at Link's car and the blood and torn clothes on the street, and at Kern's house. The way it was, an eyewitness reported, young Aller had attacked the Kern girl in the street and had been discovered by VanZuuk. They'd had a terrible battle, which didn't end until Kern arrived to save the old policeman. Which proved what everybody thought. That VanZuuk was too old and fat to enforce the law in Dellville any more, with the new criminal element they had.

Inside the house, Doyle lay shaking in his bed, afraid he would get blamed for it all. Darlene was still hysterical, sobbing and laughing, giving little curving, meaningless screams that chilled her mother's heart.

"There, there, dear" Mrs. Kern repeated over and over in a soothing tone, "I'm sure your father meant well."

CHAPTER 10

In the morning, Arnie VanZuuk paid a serious call on Mayor Travis. The old man was filled with restrained anger.

"This new policeman, Mayor . . . he's got to go. We ain't got no place in Dellville for men like that, I tell you. Ach! What he did to that poor boy."

"Well, now, Arnie," the mayor said, "you know there's two sides to every question."

"There ain't two sides to killing somebody."

"Now, now, Arnie. He didn't kill the boy. He may have roughed him up a bit more than necessary . . ."

Arnie snorted. "You seen the boy, Mayor? I tell you, that Kern has to go. I don't want him on my force no more. We don't need his kind here."

Mayor Travis patted the top of his bald head with a folded white handkerchief.

"This wasn't an ordinary case, Arnie. It was the man's daughter. And a father's feelings . . ."

"A father don't have the right to take the law in his own hands. A father calls a policeman. And if the father is a policeman, he must behave like a policeman, not a father. On my police force, Mayor, even the policeman doesn't take into his own hands the law. Ya, I tell you that."

"Uh . . . yes," the mayor said. "I see your point, Arnie. A good point, too. But we have to face reality. This Aller boy isn't what you would call a credit to the community. Always been in some kind of trouble or other. Bad family background . . . He's given us all a lot of trouble in the past, and I'm inclined to think a little lesson like this might . . . ah . . . straighten him out. He needed a firm hand before this, Arnie." The mayor stared accusingly at VanZuuk.

"Firm hand . . . pah!" Arnie grunted.

"You can't deny he's been a problem . . ."

"He's been a problem all right," Arnie said. "Our problem, and we ain't done a thing for him but pound him on the head."

"What could we do, man?" Mayor Travis cried, exasperated.

"I don't know," Arnie said, his big face creased into deep, thoughtful wrinkles. "Lots of times I looked on him and wondered what was the answer. A boy alone. Too young for the men, too old for the boys. Where's his place? What do we do for him besides sit in the police car and wait until he makes a mistake and arrest him? Eh? What do we do?"

"What can you do with a hellion like Aller? You tell me."

"Maybe," Arnie said thoughtfully, "we ask him sometime to the Rotary, or the Lions, or the Kiwanis, or the Chamber of Commerce. We got 'em all here."

"Arnie! That boy at the Chamber of Commerce meeting?"

"Why not? Maybe it does the boy good to be treated like a young man, not a kid. Maybe he eats dinner next to the banker, eh? And he is on a committee to do something in the town with the druggist or the doctor. Helps with them to build up a better town. He ain't gonna tear down the town when it's nice to him and he's on a committee. Maybe it does a boy like that good to sit down and smoke a cigar with a lawyer. Or you, Mayor. Maybe the bad boy becomes a good man. But we can't sit around and hope he comes to us. We got to go get him, Mayor. We got to go get him and make him a part of the town. Not knock him on the head and break his face for a lesson."

"You are getting old, Arnie," the mayor said. "Talking like that. I can just see that boy at one of our dinners. He doesn't belong there, Arnie. We don't have any place for him—or the likes of him."

"Then we better start to make such a place," Arnie roared. "Or we make bigger places in jails. Yah. That's what I think. Bad people don't go to jail, always. Good people send them there, you understand?"

"All right," the mayor said. "You can bring him to the next Chamber dinner as your guest. But you'll see how miserable he'll be there."

"If you make him so," Arnie grunted. "But that ain't my main business. Mainly, I tell you. Kern is out. Fired."

"The council has to do that, Arnie."

"Let it. We got no place for clubbers in Dellville."

"I don't think the council will see eye to eye with you, Arnie. A lot of people think Virgil Kern has brought this town the kind of peace officer it has been . . . it needs."

"A lot of people can be wrong," Arnie said, blinking his bright blue eyes determinedly.

"As a matter of fact, Arnie," Travis said, "the general opinion around here is that Virgil Kern is just the man we want to take over when you retire next year."

"Me, retire?" Arnie's huge body shook with laughter. "I got ten years in me yet, Travis."

"We seem to feel that next year you'll have earned your rest, Arnie." The mayor frowned. Wouldn't Arnie ever catch on? Would he have to tell him to his face?

Arnie regarded the mayor with a puzzled frown. "You mean I got to retire next year?"

"Well . . . yes. I'm sorry, Arnie."

"So be it," Arnie said with a grunt. "But meanwhile, I am the chief of the force, and I want Kern fired. We find a man to take my place who ain't so handy with the club. This fellow . . . I won't have him. No."

"I . . . ah . . . wouldn't make a big thing out of this, Arnie," the mayor said, drawing little figures on a pad. "Too many people feel that you should be allowed to retire this year. They admire the new man's ways, Arnie. I think if the council were forced to make a choice . . . damn it, Arnie, you know what I'm trying to say."

"Ya," Arnie said, rocking slightly in his chair. "I know. Arnie VanZuuk gets a kick in the behind that looks like a watch with pretty writing on it. And my nice town is run by head-knockers."

"We have to keep up with the times, Arnie. Modern civilization calls for modern methods. We're not a backward little country town now. We're growing. Ten new families moved here in the past two years, Arnie. We can't shut our eyes to trends. This town is an ideal spot for a factory or two. A river, railroads, good highway connections with the outside world . . ."

"And a policeman with a flat belly and a fat club," Arnie said. "A regular Chicago."

Arnie pushed himself from his chair and shuffled out. The mayor looked after him, shrugged, and fished in his pocket for a cigar. Invite Link Aller to the Chamber of Commerce dinners? Or to join the service clubs? VanZuuk was getting as soft in the head as he was in the belly. But the old man knew where he stood, though. There wouldn't be much more out of him. And how in blazes had he found out he was getting a watch for a retirement present? Somebody must have blabbed.

— — —

For a while Link tried to sit in the living room, but he couldn't stand it. He was sitting at one end of the couch, with his hands folded on his lap, listening to some country music on the radio. The songs were mostly about women, and every one made him think of Darlene. He hadn't seen her since that night.

His father was sitting in the chair that faced the couch. The old man was half of everything, as usual. Half-dressed, half-shaved, half-drunk. He sat back in the chair with a mocking expression on his stubby face, waving a beer can in time to the music. They were listening to Webb Pierce sing "There Stands the Glass." Whenever he came to the line, "It's my first one today," Link's father giggled and took a drink.

Link's mother, a worn, docile, but sweet-faced woman, was sitting near a lamp, reading her Bible, an expression of patient suffering on her face.

"Firrrrst one todayyyyy," Link's father chuckled slyly. He drained his can of beer, belched, and wiped his lips. "Care for a beer, son?" He grinned.

"Don't you dare give that boy alcohol," Mrs. Aller said.

Mr. Aller got up to go to the icebox, his suspenders trailing. "Well, now," he said in an amazed tone, "I plumb forgot, Maw. That is just a boy, ain't it? Now where'd I git the idea it was a big man a settin' there? One that eats policemen for his dinner?"

"You're a father to be proud of," Mrs. Aller quavered. "Makin' jokes over your son bein' beat almost to his death."

"He's a fighter, Maw," Mr. Aller said, rocking on his feet. "Don't you remember? He's the big, tough, hard guy who figgers he can whip anybody in the world because he once't knocked down his pa." He stood in front of Link. "How do you like fightin' now, Mister Aller? You still think you can lick anybody in the world?"

"Leave him alone," Mrs. Aller said. "Go get your beer and leave him alone."

"You see what happens to boys who hit their pa," Mr. Aller said, wagging his finger in Link's face. "Somebody allus gets back at 'em for what they done. You cain't go agin nature. You want to fight me now, Link? I'll fight you now."

"Leave him alone! You're drunk," Mrs. Aller said in disgust.

"Let's keep the record straight, Maw," Aller said gravely. "I am not drunk. I am getting drunk, but I am not at this moment, as I stand before you, drunk. I am getting drunk. Please pardon me while I step to the icebox to see what little treasure awaits me." He snickered at Link and went out to the icebox. They could hear him singing "It's my firrrrst one tooodayyyyyy" and laughing.

"Do you hurt anywhere, Link?" Mrs. Aller asked.

"Naw." It was hard for him to talk.

"You're so bandaged up. It breaks my heart to look at your poor face. God will punish that man for what he did to you."

"I suppose," Link said.

"But you must remember, Link, that what he did might have been God's punishment for you, for all your sins. You have sinned, Link. This ought to be a warning. I know that's why the Lord let it happen, to warn you. Link . . . Link . . . why don't you try to be godly before it's too late? The Lord will help you."

Aller came back in with a fresh can of beer. He had to stop near Link and examine his son's bandaged face again. The broken nose, the split cheek, the discolored eyes, the bruised jaw.

"My, my, my," Aller breathed. "Knowing what a daddy killer you are, I'd be afraid to see what the other feller looks like."

"Link . . ." his mother said hopefully, mournfully.

Link stood up and limped out of the room. It was still difficult to put any weight on the foot Kern had jammed with his heel even though the fractures had been set and there was a piece of steel set in the bottom of the cast to walk on.

"Careful who you pick a fight with," his father called.

"Pray, Link. Pray for guidance," his mother cried.

He went out to the garage, flicked on the yard light and sat down on the box next to his yellow convertible. He hurt, and he hurt bad. His body hurt, and his insides hurt, and his mind hurt. He didn't feel like fighting, or praying, or drinking. He felt like crying. He looked at his car, and away again. "I want Darlene," he said in a low, almost petulant tone. "I want Darlene." He rested his arms on his knees and felt small and weak and alone. "I want Darlene," he whispered.

From the house he heard his father roar "It's my firrrrrrrst one todayyyyyyyy" and roar with laughter. And his mother saying something.

"Darlene," he said again, looking hopefully in the shadows. "Darlene." He wanted her to be with him so bad he did cry, fighting every sob, mad at every tear. But it was good to let the misery take over lock, stock, and barrel for a couple of minutes. He felt better when it was over. Ashamed, but better.

He heard footsteps approach. Darlene? He knew it wasn't. He knew who it was. And Arnie knew where to find him. Arnie found a box to sit on and rested himself, puffing, overdoing the sighs and grunts because he was ashamed of the news he brought. He had promised Link to get rid of Kern.

"How do you feel, Link? You look a little better."

"Feel okay's I can, I guess. He like to have killed me. Did you see the mayor?"

"Ya, I seen him."

"Did you tell him Kern had to go?"

"Ya, I told him."

"Well, what'd he say?"

"He said? He said like this: VanZuuk is getting a watch with writing that says thank you for thirty-three years of being the policeman. VanZuuk goes, not Kern."

"But Arnie! After what he did to me!"

"E-ya," Arnie said shortly. "After, in spite of, and because."

"What'll I do?" Link asked, his battered face taking on a helpless look. "If I stay here, he'll kill me."

"I think so too," Arnie said. "You ain't got a chance here, Link. Not even Arnie can take your part. They'd fire me."

"I don't want to go, Arnie. It's my home, here. It ain't right."

"It ain't right, but what you gonna do?"

"We didn't do nothing wrong that night," Link said. "You believe that, don't you?"

"Right . . . wrong . . ." Arnie said. "Who knows what it is, and when?"

"We didn't," Link said. "And we had made our plans to get married. With me driving a truck, and Darlene working as a waitress, we could make out good. She wouldn't have to work too long. Some of them driving jobs pay a hundred a week." Link sighed. "We sure had a pretty future all figured. Until he showed up. What am I gonna do now, Arnie?"

"Go, when you are well."

"Go where? I ain't ever lived any place but Dellville. It's my home."

"You can get a job here? Kern will leave you alone?"

"It ain't right," Link repeated stubbornly.

Arnie shook his head. "You and me is both through in Dellville, Link. Only I stay here. At my age . . . it doesn't matter. You are young, Link. Young. Take your beating and forget it. Ya. Forget Dellville, Kern . . . all the Kerns. Go someplace new, make a new life. Don't go there mad. Go friendly, get a job, meet some young people. Live."

"Where? Where?"

"You find a place," Arnie said.

"I don't have any money."

Arnie looked at the convertible. "I know . . . you know . . . ten boys who want your car."

"No! Arnie . . ."

"You make another one. Better. Without the memories. Better for a fresh start. Don't take anything old. Eh?"

They sat in silence for several minutes, helpless and ashamed. They had changed. Once they had been the town policeman and the town bad boy, who had maintained an understanding and a friendship. Now what were they?

Arnie wasn't the law any longer, with all its mysterious power behind him. Link looked hopelessly at Arnie. No longer the law, but a fat old man in dark clothes and rimless glasses, his belly spilling over in a huge mound between his legs. A fat old man who wheezed when he breathed and gave off a smell of stale cheese. A fat old man who was ashamed because after thirty years of being somebody, he was nobody, and there was nothing he could do.

– – –

Arnie sat with his eyes half-closed. It hurt him to look at Link. The bandaged, broken boy seemed to have shrunk in size, to have withered. He had always been so bold, even when beaten. But now something had gone out of him. He was nervous, and his voice sounded shaky and tearful. Like a piece of metal with the temper gone. Sitting huddled on the box, shaking, his eyes moving nervously, gleaming in the light like the eyes of an animal. Like, with his thinness and darkness and narrowness, a wounded rodent shivering in a trap. Ach, God, he was still a boy.

Link sniffed noisily. "Guess there's no other way out, is there?"

"No, boy."

"I sure fouled myself up good in this town," Link said painfully. "Gettin' kicked out. Everybody'll know it."

"You shouldn't care. You go to a new life. Far away. Better."

"Will you help me sell the rag top?"

Arnie smiled faintly. "I already asked around. Darrell Atkins made the best offer."

"All right." Link's voice was dull, listless. "Take it tonight, Arnie. I don't want to see it around. I'll shove as soon as I feel good enough to travel."

"You're doin' right, boy. If I could have fired that one . . ."

"I know you tried," Link said. "Thanks, Arnie."

"I didn't do nothing. I couldn't. Thirty years the policeman, and I couldn't do nothing."

"Arnie..." Link's voice was low, almost fearful.

"Ya, Link?"

"Do you think what happened might be a kind of punishment on account of what happened to Ricky last year? My... makin' him race, and him gettin' killed?"

"Who knows that? Who knows, Link, what's a punishment and what's a reward, eh? And from where or who? Don't worry about it. The past is past. The future, worry about the future. Be a good boy, and try hard. Ya, Link? For Arnie? When I can't keep no more an eye on you?"

"Sure, Arnie." Link's throat was tight. He was afraid he was going to cry.

"And, Link. One thing."

"Yeah?" The word was husky.

"Go sudden, go quiet. Don't stop to say good bye."

"You mean to you?"

"To me... to her... don't stop to say good bye."

"If you say so, Arnie."

"All right, Link. I go now."

"Here's the keys. Take it easy, or it'll jump right out from under you. That clutch takes hold all at once."

"I'll be careful, Link. You too. Goodbye, my boy. God bless you, eh?"

"You too, Arnie."

Link stood under the yard light, a swarm of bugs swirling above his head, his shadow long on the ground. Arnie backed the convertible out of the driveway, over the shadow, and drove away, mufflers purring.

The back door of the house slammed. "Link, you out there?" It was his father.

"Yeah."

"Thought I heard you drive away."

"Just my car. I lent it to a guy."

"You gonna keep that light on all night? It's got a million bugs down on us."

"I'll turn it off in a minute."

"What?"

"I'll turn it off in a minute."

The icebox door slammed. Mr. Aller laughed and went into the house, singing raucously, "It's my firrrrrrrrst one todayyyyyyyyyy."

Link snapped off the yard light and started toward the house. Then he stopped, backed up, and went around the place where the rag top should have been parked, as though it were still there, where it belonged. Because it always would be there. Whenever he thought of it again in his life, it would be there.

He stopped once more before he went into the house and looked at the sky. The stars were out, and the moon was climbing. Like it had that night, in the field, where they had stopped for a few minutes—and gone to sleep. He saw her closed eyes, the slightly swollen cheeks, the babyish pout of her mouth in sleep. He moved his lips slightly, as though to kiss the image. "Good night, Darlene," he said. "Good bye, Darlene."

CHAPTER 11

Link was not having it easy in the city.

The cars always looked better at night, with the lights shining on them. They all looked new, and glittered, and were ready to go. So good to look at. Even the used cars. A thousand cream puffs.

Each night it was the same. First, he went into Hyman's Book Store and read the car magazines until Mike or Abe asked him which one he wanted to buy, and he put the magazine back and went out. Then he went to Pinky's and did the same thing until he knew they were staring at him, and he left to look at cars.

At first the salesmen at the used car lots had jumped to his side and had talked eagerly about this car's performance and that, and they didn't mind when he opened doors and sat in the seats. But when he came back night after night, and they knew he was just talk, they avoided him and looked angrily in his direction if he opened a door to look in or sat in a car.

Each night he made the same tour, staring in windows, wishing for this new car, not wanting that one for a gift. Each night going slowly past the lots, standing for a long, silent time in front of the sporty cars, staring dumbly at the convertibles that reminded him of the one he once owned.

Wondering when he was ever going to get a job.

There was one place where he was welcomed. A small lot on Beech Street, sandwiched in between a couple of large buildings. At the back there was a low building that contained an office, and behind that, garage space.

It was a quiet, half-hidden lot, with never more than a handful of cars waiting to be sold. The cars on the lot were usually old junkers, but sometimes there were some new cars inside the garage. But they were never there for very long.

The salesman was a young fellow named Fred Simmons. He was in his twenties, with a shining, schoolboy face and blond wavy hair. He sat around in good clothes with his feet on the desk, smoking a pipe, and liked to talk about rods and custom cars, and Link always felt welcome there.

Link sighed over a yellow Olds 88 hardtop with a black top and went to find Fred.

Fred was in the lot office, reading a detective story. As usual, his feet were on the desk and the pipe was in his mouth. He didn't look up as Link entered. Link sat down on a straight chair, lit a cigarette, and stared out the window at the street.

It was quiet. It was always quiet. Link had seen Fred sit and read and pay no attention to people looking at the cars outside. If they wanted to buy a car, they had to come in and see him. He wasn't like the usual salesman who ran out showing all his teeth like a watchdog the moment a prospect set foot on the property. He waited, and they came to him, and if they really wanted the car, he sold it to them. That waiting technique must have been a good one, because he dressed well and had money. And he was a nice guy, Link thought. The only person he'd met in Des Moines who had time to talk, and was friendly, and had even tried to steer Link toward some jobs. But they hadn't panned out.

"You get that job yet?" Fred asked, still reading.

"Naw." Link gestured angrily. "I been to see 'em all. It's the same old story. The minute them hauling companies get a look at my driving record, they won't talk to me."

"They're like that," Fred said, turning a page. "You got to be clean to drive a truck nowadays."

"I wish I'd known that when I was collecting them arrests and fines," Link said. "I thought it was fun then. Never thought it would keep me from getting a job. And I can drive. Good, if I have to."

"It's the insurance companies," Fred said. "They're awful touchy when it comes to who drives a truck they insure."

"I don't know what to do," Link said. "My money's running out. No job. I'd enlist, if it wasn't for my bad back. They won't take me. I cracked it once when I rolled on a turn."

"Do you have to get a driving job?" Fred asked, putting aside the book.

"That's what I do. I don't know anything else. But I know driving."

"There's jobs," Fred said.

"Not for me, I guess. Nobody wants to give a guy a chance."

"It sure is funny," Fred said. "Some companies are real particular about a fellow's driving record, and there's others that don't care what it says on the license."

"I sure ain't seen any of them companies," Link said bitterly.

"The other companies are probably particular about something else." Fred said.

"Like what?"

"Like how good can a guy really drive," Fred said. "And maybe how he'd rather drive than talk about it."

"I don't know anybody," Link said, "And I never was one to talk about my own business."

"And there's some jobs—good driving jobs," Fred said, putting the stem of a pipe in his boyish mouth, "that might get a fellow in trouble. And a fellow who was scared of cops couldn't get that job, no matter how good he could drive. Could he?" He gave Link a wide-eyed look over the pipe.

Link pointed to his nose. "You see where that's been busted? A cop done it. And here, where my cheek was busted. Same cop. Only I busted his nose first. And I had my bare hands, and he had a sap."

"Then you wouldn't have any love for cops, would you?" A simple question, casually asked.

"I hate them. If you knew what all's been done to me by cops..."

Fred puffed on his pipe until it was going good. He blew a puff of smoke at Link. "Let's go for a ride."

"Okay," Link said. "Where to?"

"You know the road between here and Polk City?"

"I been over it," Link said. "Curves like a snake with the cramps."

Fred looked over the keys on his desk. "Which car you want to drive?"

"I like that '50 Caddy. I've never drove a Caddy."

"It's as good as any," Fred said. "Tell you what. You drive out to the edge of town with me, and we'll see how fast you can get me to Polk City and back."

"This is swell of you," Link said, getting behind the wheel of the Cadillac. "If you knew how I've been achin' to drive something..."

"Just take it easy through town," Fred said. "Then you're on your own." He settled back, looking like a high school boy stealing a puff off Dad's pipe.

The Cadillac had a lot of torque and moved well for its weight. Link tested it on a couple corners. It wasn't what you would call a cornering car. There was a lot of iron to get around, and the weight threw a hard load on the outside front wheel. But if you didn't mind it heeling, and the tires squealing, the big car would set itself and go around.

When they got on the highway, Link opened up a little. He knew the road, but he didn't know the car. He went into the many sharp turns at a moderate speed, about ten to fifteen miles an hour faster than the signs before the turns indicated as the safe speed. He turned around in Polk City and looked at Fred.

"That the best you can do?"

"I didn't want to drive your car hard."

"Don't worry about the car. Make time."

Link responded by suddenly flooring the Cadillac and rocking Fred back against the seat. He held his foot down as the big car ate up a couple of hills and dales, picking up with a rush and a roar until the speedometer needle was wavering at the hundred-mile-an-hour mark.

A flat turn was coming up. Link held the car at a hundred, knowing it had a little more juice. He sensed that Fred was hunching down a little, the pipe gripped in his teeth. Link grinned a little, but he wasn't scared. This wasn't his car. He didn't know it too well. But he'd take the chance.

The open road let Link drift to the left side as he came on the right-hand turn. It would be easy this way. He aimed the nose of the car across the turn, and when into it he used the little reserve of power he had been saving. Even with the room, and the power, the big car didn't want to take it. The fat tires let out a rising shriek of protest as they were dragged sideways across the concrete and the car drifted to the left side of the turn, dipping to the left as Link held the wheel and forced all the power the car had into the engine.

His left tires were at the edge of the road, trying to get a bite. They slid off, onto the shoulder, and weeds whipped against the side of the car with a fluttering drumfire as Link cut them down. But he held the gas pedal down, and in the end the high torque and good bite sent him shooting back on the road, rocking from side to side as the speed needle climbed over a hundred.

"Easy!" Fred exploded.

"We made it, didn't we?"

"I don't want to get killed."

"You asked for this ride," Link said. "You're getting it."

He knew his car now, and he knew the turns coming up. He threw the big car into them at sixty, seventy, eighty, sometimes easing off a trifle before he went around, sometimes hitting his brake hard just as he threw the car into the turn and relying on the powerful engine to overcome inertia and drive him forward again. He had the wheel to hold, and he was braced. Fred was slammed from side to side as Link hurled the Cadillac through its paces. He ended up leaving a pound of rubber on the last turn, which was marked for ten miles an hour, and slowing down as he headed toward the city limits on the last straight stretch before town.

"Well," Link said, "Can I drive?"

"Yeah," Fred said. "You can drive." He puffed at his pipe and said no more.

Link waited impatiently for several minutes. "Well?" he asked, looking at Fred.

"Well, what?"

Link frowned. He wasn't sure what he was supposed to do or say. He didn't want it to be the wrong thing. "You was talking about jobs before. Driving jobs for guys who could keep their mouths shut and wasn't scared of cops."

"Was I?"

"Yeah" Link said uneasily. "That's what I thought."

"Suppose there was a job like that," Fred said. "A job the cops wouldn't like if they found out. What's that to you?"

"A job," Link said.

"You're not very particular about the kind of work you do, are you?"

"I can't afford to be."

"You can drive," Fred said thoughtfully, "and you seem gutty enough. If I sent you to a guy, and he had a job, and you took it, would you know how to keep your mouth shut if you came up against the cops?"

"I told you I hate cops. I got nothing to say to 'em."

"Suppose they had something to say to you," Fred said. "And you had to go along for a while. Could you take your rap without spilling your guts?"

"I ain't a-scared of jail or prison, if that's what you mean," Link said, trying to sound rugged.

"That's exactly what I mean," Fred said.

"They got to catch me first."

"Well," Fred said, "it's the kind of thing where a guy who shoots off his mouth to the wrong people gets to take a swim in a concrete suit."

"Yeah?" Link tried to sound unconcerned.

"That's the way it is," Fred said. "But a guy who can take his lumps with a stiff chin can do all right."

"That's me," Link said. "If it's a drivin' job, I'd like a shot at it. Otherwise, I ain't asking no questions."

"It's a driving job," Fred said. "Just remember what I told you about the . . . hazards."

"A job" Link said, breathing deeply. "A driving job. You're a real buddy, Fred. A real buddy."

Fred, who looked younger than Link, nodded gravely. "You looked like a decent kid," he said. "I wanted to give you a break."

CHAPTER 12

AFTER THE BUILDUP FROM FRED, the job itself was a letdown to Link. He was taken on at the used-car lot to wash cars, clean out the trades that came in, and do a little mechanical work.

Then, from one to three times a week, he would take the train or a bus to one of the Illinois towns just across the Iowa border. When he left Des Moines, he carried with him a set of Iowa license plates and forged registration papers made out to describe a particular car.

In the Illinois town, he would rendezvous with the driver who had come down from Chicago and put the Iowa plates on the car. Then he would drive back to Des Moines at a nice legal pace, to avoid trouble.

When he brought the car back, it disappeared, and Fred had made it clear that where it went was none of Link's business. They were good cars he brought in. Near new. Most of them were stolen in and around Chicago.

Not all the cars on the lot were stolen. Many of them came and went through regular channels, and there was enough legitimate work to keep Link busy.

There was only one thing about his job that sounded exciting. If the police tried to stop him, he was to try to escape. With the car, if possible. That's why they wanted a hard driver. Just in case.

If he was caught, his orders were to tell the police that he'd found the car parked on the street with the keys in it and had stolen it to take a ride. And he was to destroy the fake registration card before he was caught. That was all, except being prepared to serve whatever time was given him for the theft.

He liked the job. He got to drive new Cadillacs and Lincolns and Chryslers. Cars he'd never even been in before. The only thing that was missing when he was tooling along the highway to Des Moines on a nice night was . . . Darlene.

Arnie's advice or no, he couldn't forget her.

He couldn't, when her face was reflected from every shiny surface in the car, when every moonlit field looked like that field. Couldn't forget her when he saw other guys out driving with their girls, just like he'd gone out with her.

He sure missed her. All the time. But mostly when he was driving alone at night, and there was nothing to do but follow his headlights and think. Think about her. Now that he had a job, he could support her if they got married. She wouldn't even have to work. He was making good money. It never occurred to him that he was breaking the law, stealing a car. It was such a peaceful job.

He brought up the subject of Darlene the next time he was alone with Fred. He'd brought a car in from Moline, and Fred was waiting, as usual, with his young face and wavy hair and the pipe in his mouth. They went out for a late cup of coffee.

"Something happened today," Link said. "I didn't like it."

"Yeah?" Fred leaned on the table, cupping the bowl of his pipe with two hands. His white shirt was open at the neck, and he looked like a young, blond schoolboy.

"The guy who takes the toll-bridge money," Link said. "He said hello."

Link slumped in a corner of the booth, his feet on the seat. His thin face was covered by a dark growth of beard, although he had shaved that morning. His eyes looked tired.

"What's wrong with him saying hello?" Fred asked. "Sometimes they do."

"He said, 'Hello there,'" Link said. "You know. Like he'd seen me before."

"And . . ."

"He's a cop too. One of these days he'll start wondering how come I'm always coming past in a different big car. Never goin' the other way!"

Fred rubbed his pipe against his nose. "We'll arrange your meets in some of the other towns. Okay?"

"They'll get wise too."

Fred looked at him sharply. "Not turning chicken, are you?"

"No," Link said toughly. "I'm just trying to be smart about it. It won't help you if I'm caught."

"You wouldn't sing . . ."

"You know me better'n that, Fred. But who'd bring the cars in?"

"Somebody," Fred said, tasting his coffee. "There's always somebody."

"I just thought," Link said. "When I get too well known to run the cars, what then?"

"Bigger and better things, maybe," Fred said.

"I was hoping I could keep my other job with you," Link said.

Fred stared. "You want to be a wash boy?"

"It's a job," Link said. "I could bring my girl up and marry her."

"A guy with guts can find good jobs," Fred said.

"I've got guts," Link said. "But I'd like to marry my girl. Do you think I could bring her up now and do it?"

"No, man. You know that. With you out of town so much, she'd catch wise."

"I suppose so," Link said. He was feeling depressed. He'd thought about her all the way back from Moline. "But I wish I could bring her. That's why I asked about the real job, when the other is over."

"You feel lonely, Link?"

"I got a gal I want to marry."

"Will wonders never cease?" Fred teased.

Link looked curiously at Fred. "Don't you have a girlfriend? Girls ought to go for that wavy hair of yours."

"Sure, I've got girlfriends." Fred said carelessly. "All over the joint."

"I mean a special one."

"That, too. My wife is in Chicago."

"*You're* married?" Link cried.

"Not so loud. Sure, I'm married. Anything wrong with that? You want to."

"It's just that you look like such a kid."

Fred closed one eye and looked at Link with the other. There was a slightly bitter look around his mouth. "I'm twenty-seven," he said, and he seemed older when he said it. "I'm one of those people who look young for a long time; then, all of a sudden . . . Bam! I'll look like an old, withered baby. That's the way it was with my father."

"Twenty-seven," Link said wonderingly. "You sure look like a kid."

"My face is my fortune," Fred said gravely. "That bashful boy routine I do at the lot has meant money in my pocket. You know how people are. Most of them think they can judge character by the way you look. Give 'em the friendly high school face with the big smile, and they figure you're some kid they can beat in a car deal. I know that, so I help 'em think it. Stutter around, act uncertain, mention the boss in a scared whisper. It makes people feel good to deal with an honest kid they think they can cheat without too much trouble. Makes 'em feel noble inside. And it helps when the cops nose around too. I do everything but put on my school sweater for them. It's hard work, though, this bubbling-youth routine. I'm a man, and I get tired having to act like a kid."

"Doesn't your wife ever come to see you?" Link said. "Don't you miss her?"

"Funny thing about her," Fred said. "She was a real knockout when I married her. Still is. But in seven years she got older, and I didn't. You know? She didn't want to go out with me because she looked older. Like going on a date with her kid brother. Women being what they are, she left me for somebody who made her look young."

"Well, I'll be darned," Link said.

"This boy face forever has its drawbacks," Fred said.

"Why don't you grow a mustache?" Link suggested.

Fred laughed. "And have people think I was wearing my father's old one?"

"I guess you do have tough time," Link said, amused.

"Always getting chivvied when I try to buy a beer. It gets old. That's why I like the rackets. It isn't so bad letting goofs having their fun when you know you're making monkeys out of them."

"I know," Link said. "It gives me a kick too, bringing in some big car, knowing the old fool who owns it is cryin' his eyes out to the cops about it. And me, Link Aller, I'm sittin' in it pretty as you please, havin' a pleasant spin, and it ain't cost me a cent. Gettin' paid for it."

"That's the old school spirit," Fred said. "You keep that, and you've got a great future." His voice got serious. "We are a pretty big outfit, Link, and a guy with tight lips and guts can go a long way. You might have to do a little time on the way, now and then, but it's worth it. The outfit's always there to take care of the right guy when he gets out."

Link looked at Fred again. For the first time he seemed to see through the boyish mask. He saw the age and guessed the years in prison. There was something spoiled about Fred. Like an apple the worms had got into, that hung on the tree looking as red and shiny on the outside as any other apple but when it fell, it broke open, and it was all rotten and maggoty inside. He sensed the maggots in Fred.

Fred tapped his white teeth with the mouthpiece of his pipe. "Two years is just a rest," he said. "Take it from me. It's not anything to be afraid of, Link. Don't let it keep you from bigger things if they come your way."

Link nodded.

"The boys in Chicago have been watching you," Fred said. "They like you."

"I'm glad they do," Link said, for something to say.

"You're on your way up. When you get up there, don't forget the guy who discovered you, eh? I had my eye on you from the first night you came by to talk cars. I had a feeling you were what the outfit needed. Put in a good word for me up there."

"I ain't there yet," Link said, grinning.

"You will be," Fred said it quietly, but almost like an order.

Thinking about it later, alone, Link felt exultant. The boys in Chicago knew about him. And had plans for him. Him, Link Aller!

He sat on his bed in his room and pounded his right fist in the other palm. He was going places. Yes sir, he was! Doggone it, he felt so good, if there was only somebody he could tell about it. That was the hard part. Not having anybody to tell anything.

He got up and paced the floor. Arnie had told him to be good and friendly, and he'd get along. He'd been good and friendly, and every trucking company in town had turned him down. He'd have starved if it wasn't for Fred and his outfit. Starve or had to take some job with no driving!

And Arnie had told him to forget Darlene. Why should he? Why? What would the boys in Chicago think of a guy who gave up his girl because an old cop told him to? The boys in Chicago would laugh him right out of the rackets.

He liked the sound of that. He was in the rackets. On his way to be a big shot, too. Guess he could decide who his girl would be.

He hunted up the postcard he'd been saving and too afraid to use. He sat down at his table and tried to compose a message. It was hard work. He didn't want to say too much or too little. Finally, he wrote, printing the message. It began without a salutation. He printed in pencil, "I am in Des Moines and have a good job. Hope you are well." And just in case the card fell into Kern's hands, he added, "General Delivery," and sent the card unsigned, addressing it to her at the Dellville Drug, and writing, at one side of the face of the card, "Personal."

She wrote back at once, a friendly, searching letter, telling him he could write to her safely at the drugstore. In his next letter he told her how much he missed her and thought about her all the time.

She wrote that she thought about him too, and said if he sent his address it would be all right, because she would burn his letters so no one could see them.

He sent his address, and she sent him a little box of cookies. He buried his face in them, kissing them, and wrote her what was in his heart. She didn't answer, but a few days later, when he went home from the parking lot to change his clothes, he opened the door to his room and saw her sitting on the bed, looking at him with a breathless, apprehensive expression on her face.

CHAPTER 13

THEY SAT ON THE EDGE OF THE BED FACING EACH OTHER, hands gripping hands convulsively, looking into each other's eyes.

"Oh, honey . . . honey . . ." Link said. "I sure missed you."

"You're not mad because I came?"

"Love you for doin' it."

"I couldn't stay any longer," Darlene said, shaking her head slowly from side to side. "I was going to come and look for you. He was so awful after . . . that night. I can't live there. Ever again."

"You won't have to," Link said, sticking out his chin. "I'll take care of you from now on."

"He made you leave," she said. "As though that could keep us apart. I showed him."

"Do you think he'll follow you?"

"He won't know where. Nobody knows."

"How about Mr. Johnson?"

"He won't tell. He's on our side."

"The people in the post office too?"

She put her hands to her mouth. "I never thought of them."

"I don't think they're supposed to tell things like that anyway," Link said. "I hope." He looked unhappy.

"You're not sorry I came?"

"Glad, honey. Only . . . we've got to be careful. If he ever caught us . . ."

"He can't do anything if we're married, Link. Then the law would be against him if he tried."

"How are we gonna get married?" Link said. "You're only seventeen. You need your parents' consent."

"Oh," she said archly, "there are ways. I've heard girls talk. You can lie about your age, or get fake letters, or have somebody else say they're your parents. Don't worry about that . . ." She hesitated. "If you still want to marry me."

He looked at her questioning, pixie-like face, the wide blue eyes and the blonde hair tied in the ponytail with a pink ribbon, her red lips in a slight, smiling pout of hope. With her hands in her lap, and her legs crossed at the ankles, like a little girl.

"I've always wanted to," Link said huskily. "I do now. If you'll have me."

"Why do you think I came, silly? I brought all my things."

They embraced, falling back on the bed, but kissing chastely, though with great tenderness. Then facing each other, their noses touching, smiling, finding it hard to believe it was real.

"I'll be a good husband to you," Link said. "I'll work hard and bring home my pay."

"Funny Rag Top," she whispered.

"Say, how's Darrell doing with my car? I hope he's taking decent care of it." Link looked worried.

"He is," she said. "Now forget your old car. I didn't come here to talk cars. That's all men think about. Cars and football."

"And baseball," he said, making a face at her, enjoying the gentle scolding.

"Your room sure was a mess when I came in," she said.

"Clothes thrown all over, dresser drawers left sticking out. You'll have to be neater than that if you want to be my husband."

"I'm not the neat type," he said, stretching his lean body happily.

"You try keeping your room as neat as you did your car and we'll get along," she said with mock severity.

He laughed and squeezed her arm.

She sat up. "When are you going to Chicago?"

"Huh?"

"You said in your letter you had some kind of chance to work in Chicago. That's one reason I wasn't afraid to come here. I figured we'd be in Chicago before my father had a chance to find us."

"Gee, honey, I don't know," he said, troubled. "It wasn't a sure thing. It was just a chance."

"Your letter said . . ."

"Maybe I laid it on too thick in the letter," he said, looking gloomy. "I don't know how long I'll have to stay here."

"Why don't you ask your boss to give you the transfer now? He'd understand, wouldn't he?"

"I guess," Link said. "Fred's a nice guy, but I don't know how he'd feel about this. Maybe I can't take a wife to Chicago."

"Well, why not? What kind of job would a man have that he couldn't have his wife along?"

"I don't know. I'm just guessing."

"It's probably the same kind of job you have now," Darlene said. "Driving cars from Chicago to somewhere, instead of from somewhere to Des Moines. Maybe I could go with you on some of the trips. Imagine us riding along in a new Cadillac. Wouldn't people stare?"

"Yeah," Link said unhappily.

"You don't sound happy about it."

"I don't want you to build any false hopes, honey. What if the insurance company won't permit a passenger?"

"You're such a killjoy," she said petulantly. "I don't think you wanted me to come here at all. I'll bet you've got some new girlfriend now, and you're trying to get rid of me. I'll go, if I'm not wanted." She turned her head away and turned up her nose defiantly.

"Honey, honey . . . you know I'm crazy about you . . ."

He put his arms around her and kissed the back of her neck. She giggled.

"We'll work things out," Link said. "The first thing is to figure out what to do with you before my landlady finds you here. If that old bat thought I had a woman in here, she'd blow her top."

"Oh, she already knows I'm here," Darlene said with a toss of her head. "She brought me to your room."

"How'd you manage that?"

"Told her the truth. Said I came to marry you."

"Even so, she wouldn't let you stay in here tonight."

"I wouldn't dream of doing that," Darlene said. "We aren't married yet, Mr. Link Aller, and don't you forget that."

"I never have," he said, grinning.

"Your landlady has a room for me," Darlene said. "And, she has a housekeeping room that we can move into when we're married, and it has all the dishes and pots and pans we'll need to begin with. However, we will need our own towels and bed linens. I brought some with me, but we'll need more."

He watched with delight as she ticked off what they had and what they needed, frowning prettily over her fingers.

"You are the one," Link chuckled, shaking his head. "I knew you were pretty, but I never figured you could do anything useful."

She cocked her head to one side, counting now her accomplishments on her fingers. "I can cook . . . I can sew . . . I can bake . . ."

"That's cookin'," Link interrupted.

"No, it's not. Cooking is cooking, and baking is . . . things like cake. Don't interrupt. I want you to know what a good wife you're getting." She checked her fingers. "Cook, bake, sew . . . I can shop . . . I can keep house . . ." She looked at him over the tips of her fingers. "And I can love my husband with all my heart."

"Well," Link said, "that's the one chore your husband will be willing to help you at."

"You nut," Darlene said fondly. "Anyway, the three days we have to wait to get married will be time for me to get our place ready."

"Three days!" Link yelled in protest. "Why three days?"

"It's the law. We have to take blood tests anyway, and that takes time. Don't you know anything about getting married?"

Link looked at her determined expression and burst into laughter. "Our . . . marriage," he gasped. "I didn't know it was our marriage. The way you been taking over, it sounded like your marriage."

"You nut," Darlene said affectionately, looking at herself in the mirror to check her hair and makeup. "I'm hungry. Aren't you going to take me somewhere to eat?"

"Take you?" Link asked. "You mean after all that talk about cookin' and bakin' you ain't got supper fixed yet? And I thought it'd be done by now. Come on, woman, rattle them skillets."

"I'll rattle skillets after I've said 'I do,'" Darlene said. "And not one minute before, Link Aller."

She turned from the mirror, and they hugged with all their might.

"Oh, honey," Link said, sounding as though he were in pain. "I've missed you so. And you came to me. You came."

"Of course, honey," she said. "I love you, don't I?"

It was that simple to her. She loved him. And it was kind of thrilling and noble to go to him after what her father had done. In all the stories and movies, the girl's place was always with the man she loved, no matter what. She would have been willing to wait until she'd finished school if they'd stayed in Dellville, but after what her father did, she couldn't stay home. She had to show her father that she would live her own life. And after the awful way her father had acted, it would serve him right for her to run off and marry Link.

— — —

Link was glad to have her with him. All his loneliness was gone, and with it a lot of his hatred and ambition. He didn't want to show anybody anything now. He didn't want to go to Chicago and be a big shot in the rackets. All he wanted was to be let alone, and have a job of driving, and live with Darlene.

It could have been done in Dellville. Maybe he could do it in Des Moines. Ask Fred to let him stop running those hot cars and just work around the lot. Or quit and find some other job. He didn't want to fight anybody or trick anybody or steal anything anybody had. He just wanted to settle down and live with his wife. Like anybody else.

After they ate, Link took Darlene to the car lot. Fred was in the shack as usual, his feet on the desk, the pipe in his mouth. He was wearing gray flannel slacks, a short-sleeved blue plaid sports shirt, Argyle socks, and tan loafers.

"Is that your boss?" Darlene whispered, looking at Fred through the window. "He looks younger than you do."

"He's a lot older," Link said, his pride wounded that Darlene might think him inferior if he was older than his boss.

"Cute, isn't he?"

Link grunted.

"Don't be jealous, honey," Darlene said, squeezing Link's hand. "You say things like that about girls, and you don't mean anything by it. Do you?"

"Just talk," Link said.

"So's mine. You know you're the only one for me."

They kissed lightly and quickly. They did whenever they could. They had to. It couldn't be helped.

They went inside, still holding hands, awkward and bashful.

Fred took the pipe out of his mouth. "What do you know," he said. "The boy has finally and at last got him a girl."

"See, honey," Link said. "I guess that proves I didn't look at any others." He turned to Fred. "Fred, this ain't just a girl. This here is Darlene Kern, I told you about her."

A slight frown creased the skin between Fred's eyes. "You send for her?" There was a rebuke in his tone.

"Didn't have to," Link said. "She just up and came by herself. We're gettin' married . . . ?" He said it hesitantly, like a question, as though asking Fred's permission. Darlene caught the tone and didn't like it. It wasn't like Link.

Fred looked at Darlene through half-lidded eyes, his long lashes almost touching his cheeks. She was nervous, and it made her more beautiful. Somehow the lights were as kind to her as they were to the smooth skins of the new automobiles. She was young, fresh, clean, with big blue eyes and a red mouth, and long blonde hair. Young, very young.

He looked at the tobacco in the bowl of his pipe, and when his eyes no longer felt strained, he looked up at the girl with the smooth, shapely legs and the smooth face and at Link, with his narrow forehead and black hair and scarred face, with the dent in the long nose where it had been broken, the bony face, the smoldering black eyes. Those two . . . together . . . she could do better, in a way.

"Why," Fred said with a bright, boyish smile, "Bless you my children, and all that."

"It's all right?" Link asked.

"Of course," Darlene said emphatically, a flash of anger in her eyes.

"It's up to you two," Fred said with a shrug, putting his hands behind his head. "All I ask is the usual privilege of kissing the bride."

"I guess it's all right, if Darlene don't mind," Link said awkwardly. Darlene blushed. Fred nodded paternally.

"Just arrange your wedding on a night you don't have to work, eh, Link?"

"I'd kind of hoped to quit that part of the job," Link said, staring at Fred. "I'd just as soon not be running out of town all the time. From now on."

"We'll talk business some other time," Fred said. "Not now. You and Darlene go out and have some fun now, Link. You two kids have a lot to talk about. You don't want to stand around gassing with me. Run along, now."

"Okay, Fred," Link said.

"Pleased to have met you," Darlene said.

"Likewise," Fred said.

He watched them walk away hand in hand, shoulders bumping, heads close together, and suddenly he was irritable and restless.

Fred got up and paced the small office, resentment making him feel ugly. Life had been pretty dull since the organization had sent him to Des Moines for a rest. There had always been something to do in Chicago. The jobs were dangerous and exciting, and there had been something daring about his whole life. But in Des Moines he was under orders to obey the law, keep out of trouble, and get a rest. He'd been a little wild on the last two or three jobs, and the boys were afraid to go out with him until he settled down again. He was resting. But it was dull. Even running the hot cars through was dull. There was no kick in knowing they were stolen. With all the book work and responsibility he had, it might just as well be a legitimate job. He needed a kick to make him feel human again. Like his old self.

Fred paced the office frustrated and furious, like a child angry with its parents, but afraid to attack them directly. He wanted to hit something, break something, destroy something, and so ease his feelings.

Remembering the look of pride and happiness on Link's face when he had introduced Darlene, Fred's anger and resentment centered on Link. It was like a direct insult to see this ignorant, ugly, small-town punk grinning happily when his betters were miserable. Why should Link be happy? Why should Link get everything handed to him on a silver platter while he, Fred, was lonely and unhappy?

If anybody deserved a beautiful girl to love him, it was him, not Link!

The contrast between his own irritable unhappiness and Link's contentment made Fred's exile seem unbearable. He had to hit and destroy, and he could hit at Link. Yes, he could do that. He couldn't hit at Chicago, but he could spoil Link's smug pleasures.

It could be done, and it would be a terrific gag. Something that would spoil their fun, and provide him with some excitement, fun, and just enough danger to make it interesting. That's what would give it the kick—the idea of danger, and of doing a wrong thing. Then they'd be the ones who were miserable, and he'd be the one who laughed.

All he had to do was find out what day they planned to get married and then arrange to have a car in Moline that Link would pick up that same night. He'd arrange a delayed pickup, so Link wouldn't be able to get back before midnight no matter how fast he drove.

Fred chuckled as he envisioned the way it would come about. There would be Link, almost two hundred miles away from his bride. He'd really be miserable. And there would be Darlene, unhappy because the biggest day in her life had been ruined.

Maybe not ruined. Darlene was a beautiful girl. It was a shame to see her paired with an ape like Link. A shame for her to be alone and unhappy on her wedding night. Why, he might just drop around and keep her company until Link returned. Show her how much better she could do without half trying. Not that he wanted to marry her. It would be much better if she was Link's wife and his girlfriend. Much better.

Fred lit his pipe and sat down with his feet on the desk. Now life was interesting again. He'd thought up a wonderful gag to play on someone, just like in the old days. He had to admit it was something he really had talent for. Real talent. When the Chicago boys heard about it they'd know he was his old self again, and maybe they'd call him home. Meanwhile, life was going to be more interesting in Des Moines.

CHAPTER 14

THEY WERE MARRIED IN THE MORNING, by a justice of the peace. Fred and the landlady were the witnesses. The ceremony took place beside a battered roll-top desk, in a dusty office. The justice droned out the required formula, reading badly, and the bride and groom answered dutifully when they were supposed to. Link was wearing a white shirt for the occasion, Darlene a white dress. She carried a tiny bouquet of wildflowers.

"You can kiss her now," the JP said, dipping his pen in an inkwell to sign the legal documents.

Fred stepped forward and put an envelope on the JP's desk. "I'll take care of it," he said to Link. Fred was wearing a light gray summer suit with a white flower in the lapel.

Link and Darlene looked at each other and shyly kissed. Then they looked at the ring he had put on her finger. They were married.

Link stepped to the desk to get the marriage license. Fred went to Darlene. "Best man's turn?" he asked quietly. She lifted her face. He didn't put his arms around her, the way a lot of people do. He bent his head until their lips touched, his eyes open, watching her expression as he kissed her slowly, in an experienced way. She jerked her head back, putting a handkerchief to her mouth, her face pink. Fred turned away casually, as though nothing had happened, and chatted with the landlady, a plump, stern woman in a silk dress and a pink straw hat.

"Guess that's it," Link said, the certificate of marriage rolled in his hand.

"Let's go, then," Fred said, "I've got our finest car outside, and the wedding lunch is on me."

They went out and got into a black Cadillac sedan. The landlady sat beside Fred, feeling very much the grand lady as she was driven through town in the same kind of car that presidents and kings rode in when she saw them in newsreels. Link and Darlene sat in the back seat, holding hands, feeling subdued, wanting to be alone.

Fred drove them to a restaurant near the airport. Inside, Darlene and the landlady went to powder their noses and freshen up. The two men remained at the table.

"Wish I could order us drinks," Fred said. "In a decent state, you could. Care for a beer? We can buy that."

"Guess not," Link said.

"A little alcohol helps," Fred said slyly. "Helps you keep up the old courage."

"Shoot, I ain't afraid," Link said.

"She might be."

Link frowned. He didn't like this kind of talk about him and Darlene. It was their own business.

"By the way," Fred said, feeling the blood throbbing in his head. "I've got a little bad news for you. I even hate to bring it up."

"Yeah?" Link toyed with a spoon, his dark, lean face impassive.

"Yeah." Fred tried to sound sympathetic and dismayed. "I got a call from Chicago early this morning. You've got to pick up a car in Moline tonight at nine o'clock." He watched Link narrowly.

"Aw, no," Link said. "Not on my weddin' day." He didn't sound upset but resentful.

"Can't be helped," Fred said. "I told them you were getting married, but you know the Chicago boys. They said the car was real hot, and they had to move it. So, I guess there's no choice."

"I ain't goin'," Link said. There was a smoldering fire in his eyes. "I don't care how hot their car is. I ain't goin' after it today."

"You can't get the Chicago boys mad, Link." There was a threat in Fred's word. "You don't argue with them. You do what they say."

"This time I don't."

"Oh yes you do. You have to. I order it. It's my neck, too."

"Then you go get it," Link said. "Let somebody else get it. I'm not the only guy that can drive."

"The Chicago boys . . ." Fred began warningly.

"To heck with the Chicago boys," Link said angrily, throwing his spoon on the table. "If they don't like it, they can get them a new boy. I'm fed up with the job anyway. I quit, if that's the way it is."

"You can't quit," Fred said. "You know? You're in too deep. We can't let you quit."

"They can fire me."

"I don't think you'll like the way the Chicago boys do that either. You know? I told you once. Cement swimming suit, Link. Don't think they can't do it."

Link stared at the tablecloth, regretting getting hooked up with Fred. It wouldn't have happened if the other companies had given him a chance. Now he was hooked and getting in deeper all the time. And dragging Darlene along. Where would it end?

"Even so," Link said stubbornly, "I ain't workin' on my wedding day and night."

"You have to!" Fred's plans for Darlene that night were tottering. "You catch the afternoon train, Link. I'm telling you. It's an order right from Chicago!"

"I won't," Link said. "For you, or Chicago, or anybody. And you can tell 'em I said so and why. I ain't afraid of Chicago. Not over this."

"You're asking for trouble, Link," Fred said in an ugly voice. "Big trouble."

"I seen big trouble before I seen you," Link said. They glared at each other, their hatred in the open. "You make trouble for me, and I'll spread that baby face of yours all over the pavement. And don't think I can't do it or won't."

"Look, boy . . ." Fred's voice was hard.

"Look yourself," Link said. "You got me into this outfit. You asked me in. And I've done all the jobs you asked me to. But I can be pushed just so far and no more. I ain't going nowhere for you. If you want that car, you know where to find it yourself."

"You're not going to last very long, talking to me like that," Fred said, his voice low and furious. "I'll see to that."

Link haw-hawed scornfully. "I've tangled with tougher than you before I was fifteen years old," he sneered.

"We tangle for keeps in my league," Fred said. "Maybe you don't catch on, boy. If you don't want your wife to be a widow, you better be on that train this afternoon. And don't think you can squeal to the cops and get away. We've got people all over, wearing all kinds of suits. You'd never know when you were talking to one of the boys. Open your mouth once and you're a dead boy."

"I ain't a squealer or a stool pigeon," Link said. "Never was. I don't need the cops to do anything for me."

"You do what I say now," Fred said. "I told you the Chicago boys have been watching you. You do a good job here, and they'll take care of you. You'll live high. But you go crossing them, and they'll take care of you that way, too. You know? Be smart."

Darlene came back to the table with the landlady. Fred stood up when they arrived, and Link followed his example. The women sat down, and Darlene looked from Link to Fred and back again.

"What's the matter?" she asked. "You two look so glum. I thought this was a wedding, not a funeral."

"It's a business deal that's rather awkward," Fred said smoothly to Darlene. "Our regional manager called. It seems there's a car on its way, and Link will have to pick it up. He'll be home by midnight, though. There's just about time to catch the afternoon train east."

"Oh, no!" Darlene cried, her face a picture of dismay. "Not today. Do you have to go, Link?"

Link spoke quickly, before Fred could answer for him. "Aw, Fred's just kidding you, honey. There is a car all right, but I don't have to go after it. Seeing that it's our wedding day, Fred's going to take my place."

Darlene gave Fred a tender smile, even forgiving him for the nasty kiss he'd given her. "That's sweet of you, Fred," she said. "It would be awful to spoil a wedding day."

"Well, I hate to do it," Fred said, looking at Link's expression of savage triumph. "But I was about to tell Link that I have to pick up another car and can't go after this one. I'm afraid he'll just have to go, or we'll both lose our jobs."

"Oh, dear," Darlene exclaimed, on the verge of tears. "This is an awful way to start a marriage!"

Link had bunched his muscles as though to spring at Fred's throat. His face turned a dark red as he watched Fred take out his tobacco pouch and slowly fill his pipe. Darlene was sobbing, not caring who saw her.

"Oh, now, you're all without your thinking caps," the landlady scolded, shaking her head in disgust. "There, there, dear, don't cry. Don't you see how good this has worked out for you?"

Darlene shook her head, her nose an inch from the table, her tears falling on the white cloth.

"Why, honey," the landlady said. "It's so simple! You're getting a honeymoon. All you have to do is ride to Moline with Link and drive back with him. Imagine, just the two of you in a lovely big car. Why, it's the most romantic thing!"

"We could, couldn't we?" Darlene cried, lifting her tear-stained face to smile happily. "That's better than just staying here."

"It's a funny kind of honeymoon," Link said, thinking of the stolen car. "But you're right. It's better than none."

Fred stared at Darlene's tears. He saw her slipping away again and was frantic. "You couldn't do that," he said hastily. "Company policy doesn't allow riders. The insurance company . . ."

"Fiddlesticks!" the landlady roared, ready to defend her romantic scheme with her life. "Fiddlesticks, I say! She can too go with him. I know people who have taken their honeymoons that way. Some even drove to California. So, I guess these young folks can drive here from Moline! The insurance company! You two go ahead and do what the others have done, and have your little honeymoon. And if the insurance company tries to make trouble, you just send them around to see me. I'll show them a dozen cases where it's been done. Maybe twenty!"

"But Mrs. . . ." Fred refused to give up, unaware of the determination a woman could master when it came to seeing that her plans for another's honeymoon were carried out.

"Don't you 'Mrs.' me, young man," she said severely.

"You just take these two to the station right now." Her expression softened into a rough coyness. "Then you and me can have our lunch without them. You know it ain't every day I get to be alone with such a good-looking young boy like you." She gave him an arch look and giggled shrilly, surprised by her own boldness but pleased.

Fred rubbed the bowl of his pipe on the side of his nose. He felt exhausted, beaten, torn apart. It wouldn't be. After all he had tried, it wouldn't be.

"Well?" the landlady demanded sharply. "Are you going to just sit there while these folks miss their train? Come on." Fred started from his agonized trance. "Y- yes, ma'am," he said stupidly. "We'll leave right now."

"Oh, waitress," the landlady said, content now that she was in complete control. "We have to take these two young people to the station, or they'll miss their train. It's their honeymoon trip, and they can't miss that."

"No, they can't," the waitress agreed, putting Darlene and Fred together with a glance. Envying Darlene briefly.

"This young man and I . . ." The landlady pointed to Fred, "will return in a few minutes."

The waitress perked up. "Yes, ma'am," she said. "I'll keep your table for you." She stared at Fred, glad the good-looking one was coming back. "I'll be glad to serve you when you come back."

Fred went out to the car ahead of the others, the keys clenched in his hand, feeling sick with rage and frustration. Someday, somehow, Link and Darlene would pay for this.

Oh, how they'd pay for it!

CHAPTER 15

THEY GOT OFF THE ROCKET IN MOLINE, and Link steered Darlene to a small diner. "You go in and get a cup of coffee and wait for me," he said. "I'll arrange about the car."

"I'd like to stay with you," she said. "It's our honeymoon."

"It's better if you don't," he said. "In case they would get funny about my carrying a passenger."

"Don't be long."

"It won't take long."

Just long enough to put on the Iowa plates and slip the forged registration form in the carrying case on the steering post. He didn't want Darlene seeing that. She thought it was an honest job. Why couldn't it be?

He went into a small garage on a backstreet. A middle-aged mechanic in dirty white coveralls was looking at the torn-down engine of a Ford with a trouble light.

"My car here yet, Tom?" Link asked.

"Ain't due in until nine," Tom said. "Didn't you know that?"

"How would I know?" Link looked over Tom's shoulder at the engine.

Tom spat. "It was your end that said it couldn't be earlier. I don't know why. Just keeps me hangin' around when I ought to be home. My old lady squawks enough about late hours as it is, without you fellas makin' it worse."

"I didn't know that," Link said.

"Must have been one of Fred's brainstorms."

"Yeah," Link said. "I guess so." He stood back from the Ford so he wouldn't get his good clothes dirty. It didn't make sense. Why had Fred insisted the car be held back until nine? Fred knew he was getting married and would be in a hurry if he had to make the trip alone. He shifted his shoulders uncomfortably. "I hear Chicago's in a rush to get this one removed. Real hot."

"So hot it probably ain't even been stole yet," Tom snickered.

Link remembered some of the looks Fred had given Darlene. No, that couldn't be. Not on their wedding night. Not Darlene. She wouldn't. Besides she hadn't even seen Fred alone, he didn't think. But Fred sure had been anxious to get rid of him. Link felt sweaty and squeezed. It was hard to breathe. What was wrong with him, anyway? Suspecting Darlene like that. He was getting worse than an old woman. He shook the thoughts out of his head and tried to concentrate on the engine Tom was working on. It was loaded.

"That's quite a mill," Link said. "Not exactly what you'd expect to find under the hood of a plain black sedan."

"It's a full-house Merc, bored and stroked to two hundred and ninety-six cubic inches," Tom said.

"I used to have a GMC two-seventy that I bored and stroked and ran with a Howard cam," Link said. "It had everything. It would give this Merc a run any day of the week."

"Not after I get done." Tom said. "I'm fixing this one up for a short job. She'll have ten-to-one heads, a three-pot manifold, and be metered for methanol and nitromethane."

"Yeah," Link argued, "but fuel ain't as practical as pump gas for street use. That's a racing setup."

"That's what she's for," Tom said. "The boys have got a close-in job lined up. They won't have to go far but they'll have to go fast. This baby is double-shocked and can corner like a cat."

Link sighed. "I'd like to get my hands on something like that someday."

"From what I hear," Tom said, "maybe you will." He patted Link on the shoulder. "You let me fix you up a nice Ford with a Lincoln mill dropped in. You can go as fast as you have to, and you won't have to worry about burnin' up or shellin' out if the chase gets long. You could run a year on that and get good gas mileage besides."

"Well," Link said, "I'll see you around nine."

"Stick around," Tom said. "Keep me company and give me a hand."

"I . . . I've got some things to do," Link said, rubbing his nose with the back of his hand.

"Found you a gal, eh?"

"Yeah, in a way." Link chuckled, as though Tom had discovered his secret.

"Keep your lip buttoned," Tom said.

"I ain't a blabbermouth," Link said.

"I know, Link," Tom said fondly. "You're a good boy. The nights you've sat here for three hours and never said one word until you was spoke to. You'll never get in trouble that way, even if you ain't much company."

Link went back to the diner and told Darlene that the car had to be serviced before they could get it.

"I know what we could do while we're waiting," she said.

"Take in a movie?"

She shook her head, wanting to make a game of it.

Link smoothed back his black hair. "Rent a room? I've got the certificate if anybody asks."

Again, she shook her head, a glow in her eyes he had never seen before.

"Well, what?"

"Walk around and look in stores," Darlene said. "And see all the pretty things we'll buy someday. Or would like to buy."

"That doesn't sound like much fun," Link said. "Just looking in stores. But if you want to . . ."

"The way to make it fun," Darlene instructed seriously, "is to pretend that we've just bought a cute little house, and we want to furnish it, and we need everything."

"That's no lie," Link chuckled. "We even need the house."

"You want a house, don't you?" She looked at him seriously, as though their future depended on his answer.

"Sure," he said, "I'd like to live in a house. Where you had your own bathroom and didn't have to race everybody down the hall to it."

"We'll furnish the bathroom, too," Darlene said.

They strolled out in search of stores. Darlene's arm was around Link's waist, her head against his shoulder. His arm was around her shoulders, his head tilted so his cheek rested on her head. In this manner they went out to look at the living-room suites, the refrigerators, the bathtubs, the bedroom sets, television sets, chairs and tables, and all the wonderful, gleaming, shining appliances that made life easy.

They sighed over it all, and Link found that it was fun to look at sofas and chairs when they might someday be his own to sit on, in his own home. And the other stuff was interesting too, even the washing machines, and the mangles, and the blankets and linens. And the tools for the home workshop. Boy oh boy, to have a nice new clean place to live, with space for power tools to make things or work with engine parts.

It was like finding a new world for Link. Home had always meant a shaky old unpainted house, one without any modern plumbing—or even inside plumbing. A place to crawl in out of the weather. It had never occurred to him that home could be a pretty place, clean and bright and fun to be in. Not that he'd never seen nice houses; he'd never thought about them as a place where he might live.

All the wonderful, comfortable things life had to offer if a guy had enough money. Things besides cars that would be fun to own.

And he could see Darlene right in the middle of all the nice things. Cooking their meals in a kitchen with white cupboards, a big white stove, and a big white refrigerator. And both of them in the living room, on these brand-new chairs—or the big couch, watching a program on a twenty-four- or twenty-seven-inch television set. And later, both of them in one of the big, pretty beds. Maybe even one with a funny canopy over it. God, Darlene would look cute tucked away in a bed like that.

And clothes! Every pretty dress seemed just made for Darlene to wear. And the nightgowns, and the clothes that were like something a movie star would wear. Darlene ought to have all that stuff she needed, as pretty as she was. And he'd like some real nice sport jackets and maybe two good suits. Even three, someday. Maybe even good clothes, too, just to sit around the house in. Gosh, you wouldn't want to live in a brand-new house with all that new pretty furniture and sit around in boots and denims!

Looking through the stores with the eyes of a married man, Link was elated by all the wonderful things life had to offer, and he fretted because they all cost so much. But they were there, and other people had them. And if other people could, why not him and Darlene?

"You folks interested in this living-room suite?" A salesman had approached as they stood arm in arm inside a store looking at a huge beige couch with a medieval hunting scene done in a kind of bas-relief and with three large down-filled cushions. It cost six hundred dollars, and there was a matching chair for over three hundred.

"We were just looking, thanks," Darlene said wistfully.

"Perhaps something less expensive?"

"Not right now," she said. "We're just looking."

"I see." The salesman was polite, tired.

"Say, mister," Link said. "Could I do something once? Would it be all right if I sat on that big couch for a second? I ain't never sat on anything near like that, and I just wondered how it would feel."

"Link!" Darlene said, embarrassed.

"It's all right," he said in a low tone. "I've got my good clothes on." But he was embarrassed and didn't know how to say what he meant. "We got married today," he said to the salesman, explaining his new clothes.

"Well," the salesman said, holding his yardstick at each end, like a fencer about to start a match. "Congratulations. Why don't you both sit on the couch? It's a very fine piece."

The salesman had meant it as a kind of tired jest, a weary cynicism. But he watched as they gravely and carefully sat on the down cushions as though they were sitting on eggs, and he saw the longing and delight in their eyes as they let down their weight and were cradled in the comfortable, luxurious lap of the prize couch.

"Wow," Link said softly. "Imagine owning anything like this. I never knew anything could sit like this."

Darlene examined the upholstering. "Yes," she said for the benefit of the salesman. "It's not only comfortable, but it seems well made, and in very good taste."

She and Link looked at each other, sighed and started to get up.

"Thanks, mister," Link said. "Someday I might buy it, if I'm lucky."

"You don't have to get up," the salesman said. "You can sit on it all you like." He indicated the rest of the store with a wave of the yardstick. "Sit in any chair you want to, kids. Sit on the beds if you want to. It is your wedding day, didn't you say? We're open until nine tonight. Make yourselves right at home if you want to."

"That's nice of you, mister," Link said. "And if we ever get a house, we'll come here for our stuff. Won't we, Darlene?"

"It's the least we can do after the nice way we've been treated when we don't even have any money," Darlene said gravely.

They wandered around the store looking, touching, and testing, since they had been invited to do so. They wanted . . . everything. But they soon returned to the deep, soft couch and sat close together on this most luxurious of all seats they had ever known. Holding hands, dreaming, making believe they were in their own home and it was their couch.

The sound of music nearby caused them to sit up straight and open their eyes and become open-mouthed with delight. The salesman had quietly turned on a giant television set right in front of them.

"Compliments of the house, kids," the salesman said.

"Our small wedding present. All the television you care to see until we close."

And so, they sat on the expensive couch as though on a cloud, and leaned back, their heads and arms touching, and watched television on a twenty-seven-inch screen that made everything as big as life. And for two hours they were in their own home, spending an evening on their wonderful couch, watching their wonderful television set, while the world went about its business outside.

Until closing time.

"This sure was fun," Link said wistfully as the lights went out in the front window.

Darlene hugged his arm. "I don't think anybody ever had a nicer wedding evening anywhere. Ever."

They said good night to the salesman and went out, walking quickly now, because it was nine o'clock, and Link had a hot car that had to be driven across the border into Iowa before the Illinois police caught up with it.

CHAPTER 16

WHEN LINK GOT BACK TO THE GARAGE, Charley Finch had already made it in with the car, the Iowa plates had been put on, and it was ready to roll.

"Jeez, Charley," Link complained. looking at the car with an expression of distaste. "Couldn't you find anything fancier?" The car was a new Lincoln Capri, beige gray, with a pale blue top, white sidewall tires, and blue leather upholstery.

"Don't think I didn't have my eyes peeled coming down here," Charley grunted. "I don't like to push these millionaire's geek wagons around any better than you. If I had my way, we'd stick to plain black sedans. But that's what Fred wanted. Something fancy."

"Fred wanted?"

"Yeah." Charley took a cigar from his breast pocket. He was a stocky man of thirty-five, with the look of an aging athlete. He wore a blue Palm Beach suit and a cocoa brown straw hat with a West Indian print band in red and black. Charley always tried to dress to look at home in the car he had to drive.

"Fred told me," Link said slowly, "that Chicago called about this car, and it had to be picked up tonight because it was real hot."

"Fred's goin' stir-nutty in Des Moines," Charley said. "He called us three days ago and ordered the car. Said he had a buyer waiting. Said you'd pick it up tonight. No earlier than nine."

"The stupid lunkhead doesn't have any order," Link said viciously. "And he knew I had a reason for not wanting to leave Des Moines tonight. Knew it when he called."

Charley turned the cigar in his mouth and looked curiously at Link's angry, battered face.

"Fred was always a great one for practical jokes," Charley said.

"Some joke! Just the night I . . ."

"You got a girl in Des Moines?" Charley asked.

"In a way," Link said. He was watching what he said now.

"Pretty?"

"Knockout."

"Young?"

"Seventeen."

"Why, that damn Fred," Charley said without rancor. "You and me both might just get into real trouble on this deal, just because Fred wants to romance your woman tonight. That crazy nut would like it that way."

Link went rigid. He tried to control his expression, but he felt the lines of hatred and fury that were hardening on his face. He tried to talk, but he couldn't get a word out. Now it made sense. Now!

"Easy, kid," Charley said. "Easy. That kind of stuff is part of the racket, you know."

"I'll kill him!" The words came out as though they had been torn out of the flesh of Link's body.

"Kid, kid," Charley said, eyeing Link closely. "You mean it's serious with this girl?"

Link nodded stiffly.

"That slimy no-good . . ." Charley said. "We ought to yank him back to Chi and put him back on banks again. That's his line, you know, knocking off banks. Likes the excitement. We're just cooling him off in Des Moines until we get some work lined up. And I know Fred. Laying around Des Moines ain't good for him. The devil finds mischief for idle hands, they say."

"I'm gonna kill him," Link said, stating it almost in a matter-of-fact tone.

"That'd be bad for the organization, a killing inside it," Charley said. "Beat him up if you want. That's all right. I never did care for Fred myself. You ask me, the guy is lookin' for trouble. He always gets excited on a job. I was on a bank job with him. Once. Never again. He was pantin' and breathing hard all over everybody. It was disgusting," Charley said. "Very disgusting. I wouldn't want to go on any more jobs with him. I don't like to be breathed on."

"I'll teach him to try for my wife!" Link said furiously. "He won't breathe at all after I get hold of him."

"Oh, no," Charley said. "Oh, no . . . no . . . no. Your wife. I can see it. You know, that would be like knocking off a bank, see? Dangerous. Sneaky, sneaky. Listen for the strange footsteps. Your . . . wife been friendly with Fred?"

"They've hardly spoke," Link said. "Never alone, I'm sure. She's only been around a week. If you want to know, we just got married today."

"How do you like that!" Charley said indignantly. "That scum! Married today and you here and him there. For two cents I'd go to Des Moines with you and kick his pearly teeth down his throat. That cold-blooded . . ." Charley snorted, almost laughing. "But he is smart. You gotta give him credit where it's due. He's just as good knocking off a bank. Wonderful at plans."

"Cripes," Link said shakily. "What he was gonna do . . ."

"Easy, kid, easy," Charley said. "Don't let it get you down. Look at it this way. In a hundred years, what difference will it make?" He felt sorry for this lean, black-haired boy with the irregular, hurt face. The face that had always looked so hard and cold. With those dents and scars that showed it wasn't a face to be fooled with. Now it looked twisted and lost and almost frightened.

Charley patted Link on the shoulder. "Look, kid," he said. "I know it's tough, but it's one of those things. Right? You go home, and when you get there, you celebrate the wedding your way. Tomorrow, knock a few teeth loose in that Fred's head, and your troubles are over. You know, kid, we've been watching you from Chi," Charley coaxed. "We like you. Pretty soon maybe we'll have some real driving for you. Like that loaded Ford. You get in that department, and you'll make plenty of dough. A hundred bucks for every buck you turn now. So, don't let that Fred's tricks get your mind off your work, hah? You drive home nice and careful and have the rest of your wedding. Don't jinx your work and ours over a girl. Even a nice one. Go home now, Link."

Link shook his head sharply, to clear it. "She's not there. She's here, with me," he said.

Charley stared, slapped his knee and roared. "And I thought Fred was tricky! You let him set it up, then you took her with you! You kids these days! A guy can't stay ahead of you."

"She cried when she thought I wasn't going to bring her."

"Say," Charley said sympathetically, "that sounds like a real nice kid. You know something? I'll bet she didn't know a thing about this deal."

"She couldn't have," Link said. "Not when I think of what she said today. And how she acted."

"It was all that Fred," Charley said. "Listen, take my advice. You tell him you got wise and beat the living daylights out of him. It'll be all right with Chicago. They don't like Fred to get out of line and lose us any good new boys. You do that, will you? A couple good ones for me."

"I'll do it," Link vowed. "I better go pick her up. She'll be getting worried."

"Sure, she will," Charley said. "She sounds like a real nice type. You be nice to her, now. Here . . ." Charley took out his wallet and handed Link a twenty. "Here's a wedding present from old bachelor Charley. Buy her something nice and useless. That's what my women like. Nice things that are useless."

"She'll probably want dishes," Link said, almost smiling.

"Now isn't that a real pal of a wife," Charley said. "I'd sure like to shake her hand. You got a real find there."

"She's just around the corner," Link said. "She . . . thinks my job is on the level."

"You keep her in nice dishes," Charley said, "and if she finds out the truth, it won't hurt so much. I won't say hello though. You know. The fewer people who get a look at my face the better. You understand. Nothing personal. After we bring you to Chi, and she's with it, I'll come around for a home-cooked meal. Okay?"

"Sure," Link said. "See you, Charley. Thanks."

Charley waved as Link drove out of the garage. "Many happy returns of the day!" Charley shouted. He watched Link go, then went to the loaded Ford and looked it over. What a night! First driving that fancy Lincoln across the state of Illinois, then having to calm down the kid on account of Fred, and now having to drive all the way back to Chicago in the Ford and then drive the boys on a job. He'd be lucky if he was in bed by morning. All this driving and running around. What did they think he was made out of, iron or something? His back was killing him now, and most of his work was still to be done.

Yeah, he thought resentfully, a lot the big wheels cared about how much sleep he lost, as long as they got theirs. A boss was a boss. Anywhere. Ask a lot of extra work, but never a squeak about extra dough for the extra work. Bosses! They wouldn't do a thing until there was a law that made them pay overtime like anybody else. Only they'd find some way around the law. Leave it to them. And then they couldn't figure out why a guy went out and became a Red. Serve them right if he did. Come the revolution, and he'd be laying around Chi running the rackets, and the bosses could find out what manual labor was like, driving hot cars down to Moline, and pushing the rods on the bank jobs.

CHAPTER 17

Darlene was standing in front of the diner with a worried look on her face when he drove up, as though she was afraid he'd forgotten her. When she recognized him in the car, she practically squealed with delight.

"It's a lovely car," she said, sitting beside him and feeling the leather seats. "It's like that couch we sat on. I feel just like a princess or something."

"Look it over," he said. "It's got everything."

"Why would anybody want to sell this car, I'd like to know."

"To get another one," Link said. "There's a lot of people with that kind of money. Maybe we'll have it too, some day."

"We've as good as got it," she said. "We're riding in the same kind of car. Turn on the radio."

"I will, in Iowa," Link said. He drove toward the bridge that spanned the Mississippi, thinking about a lot of things. Mainly what he would do to Fred. Link felt the veins stand out on his forehead. If it hadn't been for the landlady, Fred would be with Darlene this very minute!

"What's the matter, Link?"

"What? Nothing."

"You've got such a mean look on your face. And you were muttering something."

"Just thinking," he said.

"It's our wedding night, honey," she coaxed. "Think of something nice."

He thought of Fred and Darlene. Just try to think of anything pleasant.

"Honey," she said in a low voice.

"Yeah?"

"Where'd this car come from?"

"Chicago. Belonged to some executive with the Lincoln people there. They only run a car a little while before they sell it."

"I see." She sat back against the seat, biting her nails.

"Honey . . ." She sounded almost pleading. "It . . . had an Iowa license on when you drove up."

"I know," Link said. "I brought it with me from Des Moines. Remember that little package?"

Her voice was almost a whisper. "Then it ought to have dealer's plates, not a regular license. I know, Link. I've heard my father talk . . . Oh Link . . ."

"Do you think it's a stolen car?" he asked quietly.

"It must be," she said fearfully. "What if we're caught with it? Let's get out, Link. Please . . ."

"I'm just hired to drive it to Des Moines," Link said. "I'm clean. I don't know anything."

"No wonder he gave you a job when the others wouldn't. I wondered about that when it happened. Now I know. That boss of yours . . . that Fred is a car thief. I knew there was something about him I didn't like."

"Now honey, we can't prove he's a crook on account of the license," Link said. "That can probably be explained." He was defending Fred. God!

"I want you to quit this job," Darlene said firmly. "And get another."

"Nobody else will give me one. I tried."

"Any kind of job," Darlene said. She leaned forward, her eyes bright. "Don't you see? If we can get you away, and have the police investigate, and Fred is a car thief, we'd get credit for sending him to jail!"

"Then what?" Link asked with a sinking heart.

"Then my father would see that you're a hero. He'd like you if you caught a thief. I know. He'd forgive everything we've done if we uncovered a nest of car thieves. Then we could go back to Dellville and live if we wanted to. And all be friends, the way we should be." She puckered slightly.

"I get lonesome away from my family. And I don't dare let any of my old friends see me in Des Moines. I'd like to go home." The last words came out in a thin wail, and she cried a little.

"We'll have our own home, honey," Link said, staring at the road.

"I've never been away from home alone before," Darlene said in the little, frightened voice. "I want to be near my mother!"

Link's shoulders moved in a little gesture of despair. "Well," he said quietly, "'I guess you'll have to go alone, then. I can't quit, honey. And I can't go to the cops. I'm in it, too."

"Oh . . . Link!"

"Nobody else would give me a chance, honey. I took what I could get, and I guess I'm stuck with it."

"What if you're caught?" she asked in a terrified whisper.

"Oh . . . a couple years in the pen, I suppose. But I'd be taken care of when I got out. You know. . ." A note of pride came into his voice. "The big shots in Chicago like me. I was told that if I kept up the good work, I'd get in on the good stuff. Move into Chicago and make a thousand dollars a week. Think what we could do with that kind of money. We could buy anything in the world we wanted. Like that stuff we looked at tonight."

"Doing what?" she asked, her eyes and voice dull.

"Why . . . they didn't say right out. But I suppose they know."

She shivered. "You're already counting on going to prison, aren't you? And you don't care."

"I care that I'll be away from you for a while. But it's not my fault I'm in this. I tried to be honest. I tried everyplace. Nobody wanted me but . . . this outfit."

"You didn't have to get a driving job."

"I did if I could," he said. "It's my trade. It's what I do. I don't think it's right to force me out of it for no good reason. A man's got a right to work at his trade."

"Oh honey, honey," she said. "What are we going to do?"

"What we're doing," he said. "We don't have a choice anymore. This is the kind of job you can't quit."

"You could arrange to get caught and make it look like it wasn't your fault," she said. "I know that's been done a lot. And if you tell who the others are, you'll get off easy."

"I wouldn't be here if it wasn't for what cops did to me," Link said. "I ain't squealing or making their jobs easy. I'd still be on their list when it was over. Still couldn't get a job. Not even one like this. And you know what's done with squealers, don't you? They told me. The bottom of the river. Uh-uh. I ain't going that way. I'll take my chances. It's all I can do."

She was silent. He waited, but when she didn't speak, he said, "I suppose I should have told you before we got married. But you can still back out and go home if you want to. We ain't . . . all married yet."

She moved close to him, shivering, seeking his warmth and protection. "I couldn't go home now," she said. "They'd never believe we didn't do anything bad. I couldn't go if I wanted to."

"Don't you want to?"

"I wish we both could. But we can't, can we? And you're my husband now, Link. I belong with you, and I want to be with you, and I'll stay with you, if you want me."

"I do," he said. "I do want you, honey."

"A- after all," she said, trying to smile, "we married for better or for worse. That's what we said today. Didn't we?"

"Yeah, I guess we did."

"I just didn't know it was going to be so worse . . . so soon," she said, a little laugh breaking through her words.

"It's pretty bad, ain't it," Link said. But he turned to her and grinned, if a little ashamedly. "I guess we couldn't get a worse start, so things are bound to be better."

She nodded, patting his arm. "We'll find a way to make them better."

"Don't say a word, no matter what, now," Link said. "We have to stop and pay the toll. Let me do any talking."

"All right." She sounded a little scared but excited.

Link stopped the car and handed out the exact change. The policeman on duty didn't take it but looked at the car. Darlene's breathing became louder, quicker.

"Here's the toll," Link said, keeping his head turned away as far as he dared.

"Yeah," the policeman said. "I was just looking over your car. You don't see many like it."

"I guess you don't," Link said. He knew what he would do if he had to. Floor it. Get halfway across the river and jump out and go over the side. Darlene would have to do it, too. His foot trembled on the gas pedal as the policeman examined the car with a critical eye.

"Say," the policeman said, looking at Link, "I remember you."

Link's leg muscles quivered. "You do?"

"Yeah. You're the kid that's always got a new car."

"I road test for a garage." Link said. "They let me take cars when I've got a date. I'm careful with 'em."

"That's a deal," the policeman said. "Wish I could swing something like that."

"Maybe we can make it a foursome some time," Link said recklessly.

"Anytime you say."

"I better get goin'," Link said. "The boss don't like for me to get back too late."

"Sure thing. Take care now."

"I will," Link promised. He drove away, knowing that he could never use this bridge again when the same man was on duty. Never again.

"Darlene? You still with me?"

"Link . . . I was so scared . . . I'm shaking all over."

"Nothing to it. I've done it a hundred times."

"I haven't, and I'm scared."

"We're all right now," he said. "As soon as we get across the bridge, you can turn on the radio. Maybe we can get some nice music."

She looked at the clock on the dash and giggled. "Do you know what program is on now?"

"No. I never follow the programs. What's on?"

"Dah-de-dah-dah," she said heavily. "Dragnet." They both laughed—and listened to it.

They were west of Iowa City, approaching Homestead and the Amanas, where the land was heavily wooded for Iowa. The moon was out. The car purred ahead easily, and quietly. The radio played softly.

"Remember our last night?" Link said suddenly. "It was a night like this."

"We fell asleep in that field. If we hadn't . . ."

"It was nice, though. And worth it. To me."

"It is nice to be out under the stars and the moon, in the country," she said. "The air is so nice. I can't get used to that hot stale air in Des Moines."

"Our room will be hot and stuffy, too," he said. "I see a place, I think, where we can turn off."

"All right, honey."

They parked and tried to find their love where they had left it in the moonlight, but what they found now was not the same as the old. Too much had happened; too much troubled their minds. He brought hatred and she brought fear to the blue-leather marriage bed. They were in the shadow not only of trees, but of Fred, and Virgil Kern, and prison, and their senses hopped guiltily and fearfully among these shadows like the little nocturnal animals that moved warily and uneasily in the dark brush.

And though they tried to give more and take more, the rein on their senses checked also their longed-for freedom. And when it was over, they could not understand why, although they had for the first time gone all the way, they had actually traveled but a few steps along a road they had once walked almost night-long, and to a more contented exhaustion, although they had turned back before the road came to its mysterious, and supposedly glorious final turning.

Two hours later they were in Des Moines, in the city with its hot streets, its dust and vagrant newspapers and litter, its stale air. They had reached the realm of the neon sign and the traffic light. The country, the tree-lined fields were behind them. The honeymoon was over.

Link let Darlene out at the rooming house and watched her go sleepily inside. Now he could let himself remember again consciously what this night might have been to her. He tucked his chin down on his chest as he drove to the lot where Fred would be waiting to pick up the car and spirit it away to be changed over, if change was necessary.

Fred was in the shack, his feet, as usual, on the desk. Link peered in the window.

"Hail the bridegroom," Fred said as Link walked in. Fred's face looked soft and puffy. His flesh seemed to sag down from the bones. Somehow, he looked as he had once said he would, like an old baby. He saw the set look on Link's face and tried to head it off. "Say," Fred said in a tone of jovial complaint, "did I have a time with that landlady of yours. I thought she'd never get off my back. You know, that old girl . . ."

"I had a talk with Charley," Link said.

"You did? How is good old Charley?"

"We figured out who arranged this car deal tonight."

Fred's face turned a pasty color. "It was a gag, Link. It's the kind of gag the Chicago boys always play . . ."

"We figured out why, too," Link went on in the same toneless voice that was as steady and menacing as a snake poised to strike.

"Yeah?" Fred licked his lips, his eyes darting from side to side as though looking for escape, or a weapon.

"Yeah," Link said. He sat on the desk, his left side toward Fred. Link made himself smile. "For a gag. Charley said you were full of practical jokes."

Fred's feeling of relief was so obvious as to be almost pitiful. "Sure," he said. "Just a gag. No hard feelings?"

"'Course not," Link said, the fingers of his right band curling into a tight, hard fist. "I like to play 'em myself."

"That's swell," Fred said with forced heartiness. "What's your favorite? Maybe we could get together on one, eh, Link old boy?" He summoned up his most winning boyish smile for Link's benefit.

"There's one I like a lot," Link said. "I find some nice-looking fella with wavy hair and a nice smile, and I tell him I think he's been tryin' to get me out of the way so he can do some things to my wife that ain't very decent. I tell him that."

Fred pretended not to understand. "Sounds like a good beginning," he said. "Then what . . ."

He lunged out of his chair and tried to reach a tire iron that was in the corner, but Link had been waiting, knowing Fred would try something, wondering which word would trigger him off. He was expecting the move and was ready for it.

As Fred dived forward, his fingers clawing frantically for the iron, Link stepped after him and drove his right fist hard against Fred's neck, under the left ear. Fred went down on his face, his pipe clattering on the floor. The blow stunned and partially paralyzed him. Link spit on his hands as Fred stirred and tried to cry out.

"Why then," Link said, knowing Fred heard every word, "I do what the Chicago boys tell me is the way to finish the joke. I leave the joker alive just enough to wish he wasn't."

CHAPTER 18

THE NEXT MORNING, Link didn't go to work. He lay in bed and watched Darlene with his black eyes wide and thoughtful while she moved around their small room preparing breakfast.

She had set the little table by the window and was standing in front of their two-burner gas plate frying eggs. She was wearing a white chenille robe over her nightgown, and her hair, though brushed, hadn't yet been tied up in the ponytail. She was leaning on her right leg, right hip thrust out to the side, her left hand resting on her left hip; she had a spoon in her right hand, and she was languidly splashing hot grease on top of the eggs with it. That way they would be cooked on top without turning them over.

"Breakfast is about ready, honey," she said, turning her head to peek at him over her shoulder.

"I'll be there," he said, yawning. He kicked off the sheet that was covering him and sat on the edge of the bed rubbing his eyes.

He was wearing a pair of white shorts and no undershirt. His body was lean and well-muscled, not smoothly, but as though his skin covered cables and ropes of flesh. His chest was covered with a thick growth of black, curly hair, and the hair was also long and black and thick on his arms and legs, but there it lay straighter. He looked down at himself, ashamed of his hairy body, knowing how smooth and beautiful she was.

He found his pants on the floor and slipped into them quickly, then went to take a look at himself in the small mirror near the stove. His eyes looked bloodshot, and his face was gaunt under the dark stubble. He needed a haircut. He grimaced at the narrow forehead, the crooked nose, and the sharp bones in the mirror. "Ain't I something to look at in the morning," he complained.

"You're not supposed to look pretty," Darlene said, putting the eggs on a platter. "You're a man. Men aren't supposed to be pretty."

"Some are." He thought of Fred and almost added, "or were." Thinking of Fred, Link looked at the backs of his hands, where the knuckles were bruised and skinned, and there were some cuts from where he had hit the edges of teeth.

"They shouldn't be," Darlene said. "It's not natural for the man to be the pretty one. The wife would go around all the time thinking how ugly she was, and she'd feel mean and burn the toast or something, on purpose."

He took a towel and went down the hall to the bathroom to wash.

"You won't have time to shave before you eat," Darlene said as he went out.

"It'll only take a minute," he said.

"You're just like my father. Always waits until the food is on the table and then finds something to do. I don't want your first breakfast to be all cold when you eat it."

"Okay. I'll shave later."

She watched him closely while he ate, to see how he liked her cooking, and was pleased when he cleaned his plate. To Link, it was living in luxury to have his coffee cup refilled before he had to ask, and to have someone really anxious to please him. Because she liked him, not because of a tip he might leave.

He stayed at the table, drinking coffee and smoking cigarettes. He didn't know what to do about his job. After what he'd done to Fred, he didn't think he ought to go back, but he also knew he couldn't just walk out. Chicago wouldn't like that.

"You'll be late for work," Darlene said hesitantly, not wanting to be the nagging kind of wife.

"I ain't going," Link said.

"Oh?" She raised her eyebrows. "Did they give you a day off?"

"No . . . I had a set-to with Fred when I delivered the car."

"A fight?"

"Yeah, kind of."

"You don't look it."

Link gave a sour smile. "He does."

"Did he fire you?" Darlene asked hopefully. "If he did, you're free, aren't you? We could go away and start over. I don't want to stay here anyway. I'm afraid somebody I know will see me and tell my father."

"I don't know if I'm fired," Link said. "I don't know what to do about it. I can't walk out. They might take it wrong and come after me."

"What are you going to do?"

"Sit around," Link said. "And see what happens."

"It was this way with my father sometimes," Darlene said. "People he had to arrest threatening to get even. And having trouble with hoodlums. He never knew when they'd come after him when he wasn't looking."

"I never thought I'd see his side of anything," Link said, "but I do now. If he had this all the time, I can see why he was hard to get along with. You know, wondering who's got it in for you, and what they're gonna do about it. Like I'm wondering right now."

"Do you think they'll do anything to you?"

"I don't think so. Charley said it was okay to give Fred a whipping. So, no matter what Fred tries to say, Chicago knows my side too."

"What was it about?" Darlene asked curiously.

He didn't know how to tell her. "Something he tried to do."

"I never did trust him," Darlene said. "He looks sneaky."

"He's no good," Link said. "I don't think he'd ever try to mess around with you now, but if he does . . . if he even comes near you, let me know. I'll take care of him but good."

"Why, honey," Darlene said, "he never tried to get fresh with me." She remembered Fred's wedding kiss. "But he looks like the kind who would try."

"Not anymore," Link said. He got to his feet and wandered aimlessly around the room, wondering what to do, what Fred would do, what Chicago would do. Uneasy, because he was a new boy, and hardly with them, and Fred had been with them a long time. In spite of what Charley had said, they might not like it. Might want to teach him a lesson, too. Now he had to be afraid of the people he was with, as well as the cops. Afraid of everybody.

Meanwhile he'd stick close to the room. And try to figure something . . .

Darlene said. "Let's do up the dishes and straighten up the room."

"Who cares about dishes?" Link said, kissing her neck.

"I do," she said, turning to look at him seriously as she explained. "I want to be loved, honey, but not with a bunch of dirty dishes staring me in the face, and in a lot of clutter. It's nicer when things are neat."

"Okay," Link said indulgently. "I'll even help you. But you sure are funny, thinking about washing dishes. That's a woman for you."

"I'm just being practical," she said with a mischievous look in her eyes. "The work has to be done . . ." She handed him a dish to dry. "And if we do the work first, I'll get some help."

Link stayed around the house for three days, smoking, drinking coffee, looking out of the window. He felt trapped, bewildered, angry, and scared. But he didn't know what to do. He tried to forget with Darlene, but even that had its limits.

On the fourth morning, the landlady called him to the phone. Fred was on the other end. His voice sounded different. His words were blurred and indistinct. The way people talked without teeth.

"Link?"

"Yeah."

"Fred here."

He didn't know what to say. So, he said, "Hi, Fred."

"About time your honeymoon ended, and you came back to work, isn't it?"

"If you say so, Fred. I didn't know."

"Works piled up here. And we have a car coming."

"I'll be right down, Fred."

Darlene wasn't happy that the job was still his. She had hoped they would forget Link and leave him alone. She was always afraid her father would find out where they lived and come after her.

"I'll ask for a transfer right away," Link promised. "Maybe to Chicago or some other place far away. I'll talk to Charley when I see him. He'll help us. You'll see, hon. It will be all right."

He was curious to see what Fred looked like now. Fred wasn't very pretty. There were purple lumps on his face, his left eye was black, and his nose looked almost as battered as Link's. His mouth had been hurt worst. The lips were still swollen and split, and when he opened his mouth, there were two unsightly gaps where his upper and lower front teeth had been.

The two men stared at each other for a moment. "I'm sorry I stayed away too long," Link said. "I wasn't sure I was supposed to come back."

"A hell of a time for you to goof off," Fred said with difficulty. "Just when I got banged up in a car wreck."

"That's too bad," Link said. "What do you want me to do first, Fred?" He tried his best to be polite and even humble, seeing the damage he had done. And Fred didn't even seem mad about it.

"Clean out a Chevy I just took in."

"Okay," Link said.

Somehow now, in broad daylight, at the lot, he couldn't see Fred as the awful threat to Darlene. Fred was just the guy who had been his only friend, even if it was to get him in the rackets. Maybe he and Charley had been all wrong about Fred. What if he'd been wrong . . . look what he'd done to poor Fred.

"I . . . I'm sorry I got so mad the other night," Link said, scowling at the ground. "I guess that don't help much now, though."

"I was in a car wreck," Fred said.

"Thanks, Fred."

"What for?"

"Not having any hard feelings."

"No," Fred said in a husky voice. "No hard feelings at all."

"That's swell of you, Fred," Link said. "Maybe I could make it up to you. Help you pay for new teeth, or something."

Fred couldn't hold back his feelings. "You shimple rube," he said scathingly, his fists clenched. "You don't know who I am! Do you think you'd be alive now if I had anything to say about it? Look what you did to my face!" His voice rose in pitch to a tortured cry. "You hurt me, damn you! You hurt me! No hard feelings . . . !"

Fred breathed noisily through his mouth struggling to gain control of his voice. The purple welts and bruises on his face turned darker as they filled with angry blood.

"You asked for it," Link said, if that was the way it was going to be.

"Maybe you better know something," Fred said heavily, his bruised lips twisted into a caricature of a snarl. "The only reason you're still on your feet is because the organization wants you around. But they want me too, buddy. Maybe more than they want you. They know me. They don't know you."

"So what?" Link said. "I ain't afraid of you. You don't have to look for excuses for takin' it lying down."

"You haven't been around very long," Fred said. "The way you act, you won't last too long. Maybe I'll get the job when they've had enough of you. You remember it then. No hard feelings."

"It'll take a better man than you," Link bragged.

"Tough, aren't you?"

"Tough enough. You found that out."

Fred spat unsuccessfully and winced as he wiped his chin with the back of his hand. "Why you little rooster," he sneered, "you haven't sheen anything yet. Wait until you get tapped for a real job. We'll listen to how loud you'll crow when there's lead to duck. You tell me then how tough you are, sonny."

"Anytime you say," Link answered, hooking his thumbs in his belt. "Maybe I can show you the difference between a rooster and a chicken."

Fred leaned over and spat on the ground. There were streaks of blood mixed with the sputum. When he straightened up, he had control of himself again, and his voice went back to its carefully casual tone.

"All right, Link," he said, "we've had it. Take it from me, the organization won't like it if we don't get along. They don't like trouble inside the outfit. If we can't handle it, they will, and it won't be fun. You understand? We can't put the organization on the spot over personal matters. I'd like to kill you. Maybe I will someday. But as long ash the organization wants you around and wants me to work with you . . . things have to be forgotten for a while."

"That's all right with me," Link said. "Just so you don't get any ideas again."

"Or you," Fred said evenly. "You see, you've shot your bag limit with me. Next time the organization will be on my side. So don't you forget. Any chance I get to pay you back for this face . . . No hard feelings."

His wounded mouth twisted into an unpleasant smile, exposing his ragged gums. "Shee if you're ash tough ash I am when it's your turn," he said. "We'll shee if you can tip your hat to the organization and come up with a shmile when your ox ish gored. You'll have to, you know. The boys won't stand for this kind of thing twice. They want guys around who can take anything, and shtick with the outfit. Give up anything," he added, staring Link down.

Link frowned. Did that mean Fred still had his eye on Darlene? That they'd do something to her, just to see if he'd stick with the organization? And next time he'd have to hold off? Was that the test they had for him?

"Wait until you've done what some of the other boys have done," Fred said. "Wait until you've been through the mill, and put in your time in shtir, and picked up your share of cop lead, and can do what I'm doing now. Shtaying a friend to the guy you want to rub out the shlow way. You do that—and be a good organization boy after that—and then you can come around and tell me how tough you are. Huh?"

Fred leaned far over and spat on Link's boot, then stared him in the eye. Link quivered, then held himself back. He looked back as coldly as he could. "I think I'll clean out that Chevy if it's all right with you, Fred."

"Sure thing, Link. Then wash some of the newer jobs, will you?"

"Right, boss," Link said, enjoying the game.

"By the way, Link..." Fred had taken a few steps toward the shack. He turned and came back. "Where are you having lunch today?"

"Home," Link said.

"Your new wife a good cook?"

"None better."

Fred ran his hand through his hair. "How'sh about inviting the boss home for lunch? It's the way to get promoted, you know."

Fred was watching him contemptuously. "Afraid?" Fred taunted. "Tough guy..."

Link thought about what he'd told Darlene. Never to let Fred near. What would she think if he brought Fred home to lunch? Whatever she'd think, he had to do it. If Fred could forget personal feelings for the good of the organization, he could do it, too. He'd show them all that he was just as good as any of them. That nothing would make him chicken out.

"I'll call her," Link said. "And tell her to put on another plate."

"Okay," Fred said. "We're in bushiness again."

CHAPTER 19

THE NEXT TWO WEEKS WENT OFF SMOOTHLY. Link worked at his job on the lot, made his trips to pick up cars, and got along fine with Fred.

After the first talk they had had, neither of them ever mentioned the beating. As Fred healed, it was as though it had never happened, and there were times when they sat and talked the way they had in the early days, when Link was looking for a job and Fred was his friend. The bruises faded from Fred's face, but he didn't look the same. The schoolboy look had been hammered into something else. Perhaps the change in the nose or the mouth did it. He looked more like a man, less like a cheerleader.

One day he came in with an uncomfortable, tight-lipped smile on his face. He stood in front of Link and grinned, showing his teeth. They were all there, white and even.

"How you like?" he asked Link, showing them off.

"They look better than the others," Link said.

"They ought to. The others were free. These cost me two hundred bucks. Uppers and lowers. I can take them out, too. See?" He reached in his mouth and took out the dentures, unaware of how he looked with a toothless smile. "Take a look."

"They sure do look real," Link said, examining the dentures.

Fred slipped them back into his mouth, and once again he had a full, bright, and youthful smile. "Science sure is wonderful," he said, putting his pipe in his mouth and talking around it. "Look, I've got another run coming up. Think you can take it? Two or three days away."

"Sure, I can take it. Why not?"

"You said your wife was sick. I thought if she wasn't feeling good, I'd take it for you."

It was said nice enough, but Link knew there was a challenge involved. Which came first? Personal or business? The organization liked to find out, now and then.

"I'll take it," Link said. "Maybe she'll feel better and want to go along. She likes the trips."

"Nothing serious, then."

"Depends on how you look at it," Link said. "It looks like our first one is on the way."

"Well, imagine that," Fred said, getting up to look at his new teeth in a mirror. "You didn't waste any time. When's it coming?"

"Spring," Link said.

"What are you going to name it? Figured that out yet?"

"Annalou if it's a girl," Link said. "Annalou Aller sounds pretty."

"What if it's a boy?"

Link's face clouded. "I want to call it Ricky."

"Ricky Aller isn't so hot."

"He was a kid I knew," Link said. "He got killed in a road race. Nice kid, too. Just didn't know when to quit."

"A lot of people don't," Fred agreed, trying to see his smile from the side. He was very proud of his new teeth and liked to show them off. It made him seem a very pleasant fellow.

Ever since the first night, Darlene had been making the pick-up trips with Link. Since he was most likely to get stopped at a bridge, he left her on the Iowa side, to be safe. It was hard for her to wait, and she imagined all kinds of horrible things, but he always came driving safely through and got her. The trips were no longer lonely for Link, and they always had a good time. But in the back of his mind, he knew the real reason for taking her along. He didn't want to leave her in Des Moines alone as long as Fred was around. No matter how nice Fred had been since the brawl.

But if she felt too bad to ride . . . he'd have to leave her. If he drove hard, he wouldn't be gone too long. And he'd ask the landlady to keep an eye on Darlene, in case she needed anything.

"Let me know soon as you can when I go," Link said.

"First minute I know," Fred answered. He began to hum, and Link went out.

Keeping an eye on Link through the window, Fred picked up the phone and called Chicago. He took a small hand mirror from his pocket and studied his new mouth, trying to figure out the best width for his smile.

"Hamak," a deep voice said in the phone.

"This is Fred, Al," Fred said through his teeth. "I think the kid needs some seasoning."

"You after his wife again?" the deep voice asked. "I heard about it from Charley."

"I never was," Fred said. He held the mirror up and formed the word "cheese," the way models did to get a good smile.

"What?"

"I said I never was."

"What's the rush on the kid?"

"The wife is ruining him," Fred said. "He's been talking about quitting and finding another job. It's her. She wants to live in a small town. You know the type. Strictly from the country, but the kid's crazy about her."

"Anything else?"

"Yeah." Fred bared his lower teeth. "They're going to have a kid."

"That's not so good."

"That's why I figured the time was now. If he's ever going to be any good to us, we've got to let him bake a little. If we could do it now, he'd be out in two years. I was. Two years would give him a chance to grow up, you know. Meet a lot of good men. By the time he got out, he wouldn't want to look her up again, and we'd have a guy we could use."

"Sounds logical," Hamak said. "You think he's worth all that trouble to us? It's a lot of bother getting a guy put away if he ain't worth it."

"He'll be as good as gold," Fred said. "Everything it takes once we get him pried loose from that woman. Tough, lots of guts, and drives like his mother was a V-eight."

Hamak chuckled. "If you say so, I'll take your word for it. Although Charley speaks very highly of the boy. In two years, we'll need a man to take over some of Charley's jobs. He's slowing down."

"This is the boy," Fred said, grinning at the ceiling. "When can you do it?"

"How about Thursday?"

"Fine. My side or your side?"

"My side," Fred said. "The more trouble we keep out of Illinois the better."

"You're a smart boy."

"Al . . . when do I come back?" Fred's voice sounded a little strained. "I'm going nuts stuck out here. When do I get back in my line?"

"When I say so."

"Can you make it soon, Al? A month or two?"

"I'll let you know."

"Al..."

The receiver clicked at the other end. Fred sighed and hung up. He lifted his hand mirror and practiced a sneer. But when he thought of Link's future, it changed to a smile.

It was simple. They gave him a car with the original license plates and registration still on it. Then tipped off the Iowa police. And Link wouldn't talk, even if he wanted to. He knew what would happen if he talked. And there was no other place to go after he got out. He needed them. No matter what happened Thursday night. The first time, he'd been going to do it as a gag, but Link had changed that. It would be sweet revenge this time.

CHAPTER 20

"I WISH I COULD GO WITH YOU," Darlene said in a small voice. She was lying on her bed, her hair brushed down long and full, her eyes seeming very large, and her face small and childlike.

"I wish I didn't have to go," Link said. "I wish . . . What good's wishing . . . I'll burn up the road coming back. It won't be long."

"Maybe I could go. I might feel all right." She sat up and then leaned back, looking pale. "No," she gasped. "It's the motion that does it." Beads of sweat stood out on her forehead. She looked helplessly at Link and put her hand to her mouth. "I feel so sick," she moaned. "So si—" She leaned over the edge of the bed where a pan was on the floor, retched emptily, and fell back, her breathing loud and rasping.

"Don't go," she panted. "Please don't go."

Link sat down on the bed and took her clammy hand in his own. Seeing her like this frightened him. He didn't know she would feel so bad so soon. He thought the bad part was at the end. "I have to go, honey. There's nobody else."

"Fred could . . ."

"He can't," Link said, knowing Fred could. Knowing he had to prove that nothing made him chicken out on a job. So, he'd be with them and move up to better things. If he could just quit it, like any other job! But he couldn't quit. There was no way to go but in, deeper and deeper. And with a baby coming on. There had to be some way out. There had to be!

"Honey . . ." Her eyes were closed, and she seemed to be resting better. "Be careful."

"I always am," he said. "You know that. I'll ask the landlady to look in on you."

"I'll feel better later," she said. "I always do. I'll probably be all right when you come home."

"Sure," he said, pressing her hand. "It's just an attack." He bent over and kissed her lightly on the lips, and she murmured something. He went quietly to the door and looked back. Her breathing was easier. She seemed to be sleeping. She was a pretty thing. Too bad she had to feel sick. He closed the door softly after himself and tiptoed down the stairs.

— — —

In the evening, Fred took a long, hot shower and dried himself with a rough towel to make his smooth skin glow. He combed his hair carefully, and taking out his new teeth, he held them in his hand and gave them a thorough brushing. He put them back in his mouth and tried out his smile. He took the teeth out again and, gave them a second brushing.

He inserted his teeth in his mouth and clicked them together. Then he dressed, putting on a pair of pale blue nylon shorts and a nylon undershirt. He chose his gray flannel slacks, a navy blue sport shirt, and a lightweight tie-scarf of pale blue with navy polka dots. Navy socks and cordovan loafers. He posed before a full-length mirror and smiled a provocative smile at himself. "Oh, you handsome devil, you," he said to the image. "What wicked deed are you planning tonight, may I ask? As if I didn't know!"

His telephone rang. It was Charley, in Moline.

"He's on his way," Charley said. "And they're waiting for him."

"How's the boy?" Fred asked. "In good health?"

"Worried about his wife," Charley said. "She ain't feeling so good."

"So I heard," Fred said. "But it's nothing serious. I'm sure she'll feel much better in a little while." He looked at the mirror and said, "Cheese," and was rewarded with a dashing smile. Humming softly to himself, he put his wallet, change, keys, and handkerchief into the right pockets, put his pipe in his mouth and struck a serious pose. Hard to believe that handsome youth in the mirror had served time in prison, knifed two men, beat up several women, robbed a score of banks, wounded at least one policeman, and had been shot in the leg himself.

He stood, lost in thought before his smart image. There was something different about his face. It was the same, yet not the same. There was an older, more serious expression on it. And the features seemed more mature. In a way, it was what he had always wanted. Link's fists had aged him just enough to keep him from looking like such a kid. "You," Fred said gravely to his image, and getting the words back, "are no longer a good-looking, charming, juvenile delinquent." He straightened up and looked stern. "You are a mature hoodlum."

— — —

"There she is," Charley said. "All yours."

"A two-year-old Plymouth," Link said in disgust. "Just when I wanted to make time."

"You'll get there," Charley said. For the Plymouth, he had worn slacks and a sport shirt. No hat. His hair was brown, straight, and sparse. The bald area usually covered by his hat looked pale green in the light against the red of his face. He chewed on his cigar and belched slightly. His stomach couldn't take this driving all the time and the irregular meals. And this deal with Link.

"I'm in a hurry," Link said. "My wife is sick."

"You shouldn't have come, boy," Charley said. "I always say a sick wife is more important than a hot Plymouth two years old. Fred could have come."

"I don't want the guys in Chicago to think I can't be counted on," Link said. He looked haggard under the bare light bulb hanging from the ceiling. "No matter what."

"Listen," Charley said, "the Chicago boys are human, too. They know what it is to have a wife or a kid get sick just when there's a bank job or a load of furs to heist. That's life. You gotta expect those things. That's why I'm a bachelor. This life ain't for a married man with kids. It's always harder for them. They got worries us single fellas never heard of. I feel sorry for them, believe me. My heart's bled for more than one guy I've driven with on a job in the middle of the night who had a sick kid at home. Trying to keep his mind on his work when he'd rather be home taking care of his kid than shooting it out with night watchmen and cops. No wonder a guy like that misses. How can he concentrate? Heh?"

Charley rolled his cigar in his mouth and laughed. Link had to smile. You never could tell when Charley was serious or kidding you. He could do both at once, somehow. At first Link had believed every word, until he'd mentioned some of it to Fred, and had been laughed out of the shack.

"Fred didn't give me no kit," Link said. He was referring to the Iowa plates and the forged registration certificate.

"It ain't worth wasting plates on this one," Charley said. "We stuck on some fake Illinois. But hell, who's gonna bother you driving this heap?"

"Yeah," Link said, trying to get in the spirit of the thing. "Who'd think anybody would steal on old Plymouth? Not even a cop would think that." Somehow his joke didn't come out as funny as Charley's, but Charley laughed.

"You know, Charley," Link said, "my wife's old man is a . . . against our being married, and he's making it tough for us. Do you think I could get transferred somewhere, so we could get away from him? You know how it is."

"That in-law problem," Charley said, shaking his head. "It's ruined more good mobsters. A guy goes home after a hard day in the rackets, and there's the mother-in-law wanting to know why her daughter don't get as good as her friend's daughter. I've heard it. 'For strangers you steal Cadillacs, but for your own flesh and blood you steal a Ford.' And wanting to know why you can't buy your wife a clothes dryer with your share of the loot when the other guy on the job bought his wife a dryer and a fake fireplace for the living room. I'm telling you . . . in-laws . . ."

"I know, Charley," Link said, a little impatient. "But this is rougher than you think. If he snooped around, he might cause trouble. Couldn't you ask the Chicago boys . . ."

"To rub him out for you? That's a big favor when the fella has to travel all the way to Des Moines."

" . . . if I can have a transfer? Some nice smaller city, maybe. We might be able to buy a house in a small city. With the baby coming . . ." Link rubbed his head with an air of desperation. "If you could, Charley."

"I'll do what I can, kid," Charley said, shaking his head in a slight, pitying gesture. "I'll get right on the phone."

"Thanks a lot, Charley. Any time I can do anything for you . . ."

"Buck up," Charley said roughly. "Quit crawling. You don't have to beg me. Go on home to your wife."

"Sure, Charley. Be seeing you." Link hopped in the Plymouth and drove it out of the garage and headed toward home. Darlene would be glad to hear there was a chance for a transfer. Make her feel better, maybe. Wonder how she was, all alone and sick. Of all the nights to have a used Plymouth. When he needed a fast car to get home in.

- - -

Charley went out to the diner where Darlene had waited for Link on their first trip. "Glass of milk," Charley said to the cook as he walked in. "A short one. Not too big a head." He went back to the pay phone, looked up the number of the Davenport, Iowa, police across the river, and dialed their number. "I want to report a stolen car," Charley said. "It's a Plymouth, two years old, black tudor." He waited, then gave the license number and the name on the registration card. "I'm Jerry Reynolds," he said. "The owner. The car was stolen here in Moline a couple minutes ago. I was in a restaurant, eating. A carload of fellas stopped by it, and one guy got out and jumped in my car and drove away in it. No, I don't think he knows I saw him. When I ran out, I saw the other car had an Iowa license, so I figured he would be heading your way . . . That's all right, officer. What? No, you don't have to bother with that. I'll call you . . ."

Then he called Fred and reported all was well.

Charley hung up, put his cigar in his mouth and started out. "Hey," the counterman said, "your milk."

"Feed it to the cat," Charley said, tossing a dime on the counter.

"We don't have a cat," the counterman said.

"You don't! How do you keep the rats and mice out?"

The counterman shrugged. "Don't have a cat, and don't have a single solitary rat or mouse or cockroach in the place."

"No cockroaches," Charley said hollowly. "And you dare to call yourself a restaurant." He went out shaking his head.

The counterman looked after Charley's retreating figure. "A nut," the counterman said aloud to himself. "A guy like that has to come in here. I better look and see what he did to the telephone."

Charley stopped outside to relight his cigar, which had gone dead. "That kid gets out," he said to himself, "he'll thank me if he finds out, if he doesn't kill me. What's he need with family in this profession? He'll thank me in two years. With tears in his eyes. He's gotta learn, that's all. Like everybody else. Two years in stir is just what he needs now. Make a man out of him. Two years. What's two years? His age, he can do it standing on his head drinking a glass of water. Be better for his wife to go back to her family and have the baby anyway. If she only knew it. She could marry a square John, and the kid wouldn't have to be carrying his report card to the penitentiary so his old man could sign it. No place for a school kid to hang around anyway, the penitentiary."

Charley sighed. They'd all be better off this way, but they were probably too dumb to know it. Just like he was too dumb to understand what a good thing he'd done. And then they wondered why a guy chucked it all and joined the Salvation Army!

— — —

He'd been pushing the Plymouth through Davenport, but somehow, he wasn't alarmed when he noticed a police car behind him. Police cars were all over, and who'd be looking for the Plymouth to be a hot car? Not even the police.

He held to the legal limit, though, so he wouldn't attract attention. He wished they'd turn off, so he could speed up. He wanted to get home!

He'd been followed before, in his day, and he knew the police were following him. Just by the way the cruiser was creeping up on him, slowly, quietly, as though by accident. Why now, tonight, when he had a slow car and Illinois plates on!

Now, when he needed a Lincoln, or a Chrysler, or a Caddie, or a Hudson, or an Olds, or anything fast. Now he was stuck with the Plymouth.

He couldn't outrun the cops, but maybe he could outthink them. Story . . . story . . . his wife sick . . . that could be verified.

He was sweating, and his hands were watering the steering wheel.

Sick wife . . . sick wife . . . had to get home. Borrowed car from friend that might do it . . . not so strange . . . Jesus God, what a night to get tailed. There wasn't any story. There could have been, with Iowa plates. Not with Illinois. Why, how?

The prowl car was getting closer. He knew what it would do. Get in position and then suddenly cut in front of him and block his escape. He saw a dark street to the right and cut into it, resisting an impulse to speed. With a black car, on dark streets. He could cut his lights. Once out of sight . . . gone. It wouldn't be the first time.

The police car took the corner. That proved it. They were after him. He didn't hurry. Didn't want to make them jump him yet. He slowed, as though looking for a house, then spurted into the next corner and cut his lights, his foot to the floor. He took the corners as they came, whichever way he could, flitting past parked cars with only inches to spare, his eyes straining to see ahead in the darkness.

The police car seemed to know where he was going. It came after him, using its searchlight to pick him up. He wasn't fast enough to get two corners between himself and the cops and out of the light's range.

Now he began to work his way toward the west edge of town. If he could just get a jump. Enough to give him time to jump out of the car and get lost. Where there would be some place to hide. Some cover. They were too close. By the time he stopped, they'd have him in the light and be able to follow him on foot. That's when they started shooting, too. When you ran from the car.

If he could make it to the highway! If they tried to pass him, he'd spin them off the road. He'd have to. They were probably calling for other cars now. Boxing him in. What to do? Darlene . . . Oh God, Darlene, help me . . . help me . . . What will happen to you if they catch me, honey? Oh, honey . . . I need help . . . help . . .

He was breathing hard, sobbing as he failed to find a way out of the trap. They had him bore-sighted. Whoever was pushing that police car could drive.

He nosed the Plymouth around a corner into a narrow street, driving blindly. For the first time, the siren went on behind him. It wasn't a new sound to his ears, but awful, awful . . .

He stood on the brakes as a bank loomed up ahead of him. He couldn't stop in time and hit the bank with his tires dragging. He went up a little way and rolled back. When he came to a stop and got his breath back, the police car had its nose against his back bumper. He was trapped. That siren had been to warn him of the dead end. Oh, God! There had to be some way out.

The policeman who drove the prowl car got out and approached holding his flashlight. One. Link turned to see what the other cop was doing. There wasn't any other cop. There had to be, but there wasn't. Somehow this one was alone.

Link sat behind the wheel and waited. The cop was big. That was the first thing he saw. Too big to fight. What could a guy do . . . One thing. Turn the door handle so that the catch was slipped. Just a fraction of an inch. Maybe. Maybe.

The light flashed in Link's face. He squinted into it, trying to look innocent. The light left him and played on the car. On the license plate. The policeman glanced at a piece of paper in his hand.

"What . . . what's the matter, officer?" Link asked. "Is my car wrecked? I guess I made a mistake trying to race with you, didn't I? I didn't know you were a police car at first, and then I got scared. You know how it is, I guess."

The policeman stopped a few feet away and put his light on Link again. Link grinned sheepishly.

"Hand over your driver's license."

"Sure, officer."

Link got out his wallet and held it out of his window a few inches.

"Bring it here."

"I don't know if I can," Link said. "I hurt my leg when I hit the bank."

The policeman came a step closer and reached for the license, bending forward slightly. And Kern's trick was turned again. Link's feet hurled the door open, hitting the policeman on the forehead. He grunted and sagged to his knees, pawing for his gun. Link was out of the door and on him like a cat. He drove his knee into the policeman's jaw, completing the knockout. Then he started to run, took two steps, and swung back.

Working quickly, he slipped the policeman's gun out of its holster and slid it into the pocket of the poplin jacket he was wearing. He took the extra rounds of ammunition in the belt. He found the policeman's wallet, with eight dollars in it, and took that. Then he handcuffed the inert policeman's wrists behind his back.

As he straightened up, the policeman groaned and twitched his legs and tried to cry out. That frightened Link. He pulled the gun out of his pocket, and flinching, hit the policeman on the side of the head two or three times. The officer lay still. Link put the pistol back in his pocket and felt something wet on his hands. Blood. It was on the pistol, his hands, his jacket.

He threw the pistol as far as he could, hurled the jacket in another direction, and ran.

He stopped running when he was on the verge of collapse. What to do . . . what to do . . . They'd be checking the bus stations and freight trains. Have to steal a car. The only safe way.

He tried three before he found one that was unlocked. There was no key in it, but he didn't need a key. All he had to do was to get at those wires under the dash, and with a coin . . .

He wasn't even sure he knew what kind of car he was driving. There was only one thing in his mind. Get to Darlene. Get Darlene and run . . . run . . . How much money did he have? Eight off the cop and about ten of his own. He felt for his wallet. It was gone. He had it in his hand and held it out to the policeman . . . and forgot to pick it up!

Go back? He didn't dare. That blood. Maybe the cop was dead. Or had been found. They'd be looking for him and know about where to look. His name, address . . . They'd radio Des Moines, and his house would be under guard long before he could get there. Where? Where?

He stopped the car, turned it around, and headed back toward Illinois. The garage. Find Tom. Find out where to go in Chicago. They had to take him in. There wasn't any other place for him to go. He'd get in touch with Darlene somehow, so she'd know. Then she could follow him.

"Hey! Hey you!"

Link cringed, as though expecting a blow. He looked out at a familiar face in a police uniform. Tall Corn Bridge. He'd crossed it in a blank and didn't know he had stopped.

"How you doing? I haven't seen you for a while!" It was the friendly bridge toll collector.

"I haven't been this way," Link said.

"When are you and me going on that big date in one of them big cars of yours?"

Link felt he was going to scream. "H- how about next Saturday?"

"I work then. How about the Tuesday after that?"

"Fine," Link said, the blank, staring look still on his face.

"I'll give you my address, so you'll know where to pick me up," the guard said. "I live in Davenport. Are you coming from the east or west?"

"B- both," Link mumbled. "Either one. Whatever you want."

The guard looked at him closely. "Are you drunk?" he asked sharply.

"No," Link said. "Tired. Tired. I've been working awful hard."

"It's all right. I thought if you were drunk, I'd have somebody drive you home to keep you out of trouble. No sense getting in trouble over a little drink, you know. Now to reach my house you come down the main street where the highway runs, and you turn here, at the second stop light. Turn north one block and then keep going until you come to a white house on the southwest corner of a three-street intersection. Take the northeast street to the east and go two blocks and turn south again."

Link's head drooped. Darlene, he was crying inside. Darlene.

There were blood spots on his shirt.

"... and that's the house. Got it?"

"Yeah," Link said. "See you Saturday." He put the car in gear and drove off. "Tuesday!" the guard yelled at him. "Not Saturday! Tuesday! Hey, wait, you didn't pay your toll!"

Shaking his head and swearing, the guard paid it for Link. He'd get it back the next time the fella came across the bridge. You couldn't call ahead and stop a friend for a little thing like that. Not a friend who had his pick of cars.

The phone rang in the guardhouse. The guard answered. It was the Davenport police wanting to know if he'd seen anything of two stolen cars. One was a black Plymouth with an Illinois license. Abandoned by the same man who then stole a 1949 Ford, dark blue sedan, after attacking a policeman.

"No," the guard said. "I've only seen one dark blue Ford in the last hour, and that was driven by a fella I know."

He hung up, looked thoughtful, and scratched his chin. "Or do I? And that was the first time I ever saw him drive into Illinois."

The guard reached for the phone. If he was wrong, he could always tell the fella he had to do it to get the toll, and there wouldn't be any hard feelings.

Link abandoned the Ford as soon as he was out of sight of the bridge and caught a bus to Moline, carefully holding his hand over the bloodstains on his shirt. When he reached the garage, on foot, it was open, and Charley was still there, sitting on a chair and looking at a newspaper.

"What the hell!" he exclaimed as Link staggered in.

Link leaned against the wall, his body shaken by sobs of fear and relief. "Oh, God, Charley," Link choked. "I had it. I had it."

"What happened, kid? Tell your Uncle Charley..."

Charley went after the pint of bourbon he always carried and made Link take a drink.

"They hopped me," Link said when he could talk. "Some crazy cop got an idea to tail me. We tangled . . . I don't know . . . I think I maybe killed him . . . I didn't have no place to go, so I came here."

"The boys will love taking in a kid who knocks off cops," Charley said. "But, maybe he ain't dead. Could that be?"

"I hit him on the head," Link said. "There was blood."

Charley looked at Link's sunken cheeks and staring eyes. A fight with a cop. Blood. The kid had had his seasoning without the two years in stir. If he could get away after the way he'a been fingered, just think of what he could do with a little training.

"Charley," Link pleaded. "Do I get to go to Chicago with you?"

"Gee, I don't know, kid," Charley said, scratching his chin. "It ain't up to me to say. I'm one of the listeners in the club, not a talker."

"What'll I do, Charley? What'll I do?"

Charley pondered the question for a moment, his round face showing a benevolent concern. "Tell you what, kid. I'll call Chi and find out. Meanwhile you stay here and help yourself to the jug. Read the paper if you want. Little Abner's real funny tonight."

Charley winked reassuringly at Link and went out to find a phone booth. He put in a call to Hamak, whistling through his teeth while the phone rang.

"Al . . ." Charley began without saying hello. "Guess what."

"Guess what, what? They take the boy in?"

"As a matter of fact," Charley said. "They didn't."

"You're a fixer, all right. What did you do wrong? It couldn't miss."

"It didn't," Charley said. "But the kid got off the hook."

"Yeah?" Hamak's voice was quiet.

"Yeah," Charley said. "He slugged the cop."

"How bad?"

Charley looked at the end of his cigar. "Maybe all the way." He waited while Hamak swore angrily.

"Where's the kid, Charley?"

"With me. He wants to come to Chicago. He ain't never seen the big city, and he thinks he's got some vacation time coming."

"Quit your kidding. Who wants a dumb kid around with a murder rap coming? A cop at that."

"Maybe the cop ain't dead," Charley said optimistically.

"What do you think, Charley? You know him."

"If I turn him loose, he'll get picked up," Charley said. "He's all upset, on account of this is his first booboo. They'd catch him easy, and he'd probably spill, the condition he's in."

"You want to take care of him?"

"That ain't so simple down here," Charley said uneasily, resenting the way Hamak was trying to unload the thing on his shoulders. If anything went wrong, there went Charley. A lot Hamak cared about that.

"Tell you what, Al," Charley said. "Let's wait and see what happens to the cop. If he lives, we don't have a thing to worry about. We can start working the kid in. He's good people, Al. He's got the stuff. It'd be a real shame to lose him if we didn't have to."

"What if the cop dies?"

Charley rubbed the side of his nose thoughtfully. "We can always use Lake Michigan if we have to." Charley's voice became more cheerful and confident. In Chicago, somebody else would handle the messy details if the cop died. "I vote for the lake, Al. From what I know, he ain't the kind of kid anybody's going to look for too hard if he shows up missing."

"Sounds all right," Hamak said. "Bring him in."

When Charley returned to the garage, Link was sitting on a chair, his arms folded across his body as though his stomach hurt. He was hunched forward, staring at the floor. "Come on, kid, cheer up," Charley said. "You look as though you lost your best friend, and you didn't even know the guy."

Link shook his head slowly, not answering.

"Come on, kid," Charley said, "you asked me for a transfer, and I got it for you. You're on your way to Chi."

Charley put the bottle of whiskey in his jacket pocket and herded Link into the car. In a few minutes they were on the highway.

"Come on . . . kid, come on," Charley coaxed, trying to rouse Link from his stupor. "Be a little company, will you? Tell you what, kid. While we're driving, you think up a nice name for yourself to use in Chi. That's one of the nice things about this profession. You're not stuck with the name you were born with if you don't like it. You're even luckier than a movie star. They only get one change. But you, kid . . . you can give yourself as many names as you want to. A new name with every suit if you want to. A different name for holidays, if you observe holidays. I'm asking you, kid, isn't that something? All the names you want, whenever you want them? Don't even have to use the same one twice, if you don't want to carry around a used name. I ask you where else but in America? Come on, give Uncle Charley a big smile, whatever your name is."

When they reached Chicago, it was Richard Monroe, which was as close to Ricky Madison as Link dared to get.

CHAPTER 21

IN THE EVENING, DARLENE FELT BETTER. In fact, she had begun to feel better right after Link had left and she didn't have to go along. She was getting tired of the same old trips to Illinois and back.

And although she had asked Link to stay, she was glad that he had gone, really.

She had bathed, put on a nightgown and her chenille robe, fixed her hair, and put on makeup. Now she sat in their one big chair, with her feet tucked up under her, waiting for Link to come home. Really wanting to go out somewhere and do something.

Did she dare?

It was kind of awful to be in Des Moines like this, a prisoner in a room. And that's what she was. A prisoner.

It had been exciting and daring to defy her father and run away to Des Moines to join Link. The way her father had treated her.

Everybody knew about it, and the whole town was laughing at her. She'd been ashamed to look the boys in the face.

And when Link wrote and said he was going to get a job in Chicago, she saw a way to get revenge on her father and get out of Dellville. She'd go to Chicago with Link. There'd be a lot of things to do in Chicago. And he had been driven out of Dellville because of her, and she did love him, and she didn't want to stay in Dellville where everybody was snickering and with her father being so mean.

Darlene put a fresh stick of gum in her mouth and chewed it down.

This wasn't any fun. Golly, the thing she had liked about Link was his spunk. Nobody could tell him anything there. He was exciting in Dellville. But he sure wasn't exciting in Des Moines. Golly, she had so many friends in Des Moines, too, and she couldn't call any of them. She and Link were afraid her friends might talk if they saw her, and her father would find out where she was. And the other policemen knew her too, and she had to hide from them. She was just a prisoner in this old dumb room. The only time she went out was once in a while with Link, when they would sneak to a drive-in movie or she would buy groceries at the corner store.

She wouldn't have cared if it had just been for a few days. That would have been exciting, hiding from everybody for a while. But it was going on, and on, and on, and they might never get to Chicago, and she'd be cooped up in this room for the rest of her life! With a baby! It was awful, being cooped up forever before you'd had a chance to have any fun!

Did she dare? Did she dare call one of her girlfriends and have her come over while Link was gone? Maybe Jean, or Pat? They wouldn't tell. And she had to talk to somebody! She couldn't stay cooped up like a prisoner in this room forever!

She'd begged Link to quit, or ask for a transfer, but he kept putting her off. He was afraid he'd make his gang mad if he bothered them. Link afraid. He sure wasn't the same Link here in Des Moines. He wasn't brave or reckless or daring or fun at all. He didn't even have his own car! He was just a car washer, and he did everything that old Fred told him to. Even left his sick wife alone, because Fred said he had to go get an old car!

She pouted and sniffed, and thought angry thoughts, and then felt so lonesome she put her head down on the arm of the chair and cried. This wasn't the way it was supposed to be at all. This was worse than being at home. At least there she'd had Link to slip around with, but now she couldn't even do that. What if she had to spend the rest of her life hiding in this little old room? Be an old woman who'd never had any fun? And Link wasn't as interested in her as he had been, either. He just did what he wanted when he wanted to and didn't think she needed any loving, now that they were married.

Why couldn't they go to Chicago! So she wouldn't have to be cooped up anymore and afraid to show her face outside the door? She couldn't stand it like this. It was so dull, so awful . . . She cried and chewed her gum and swallowed tears past it and wished she could go out with her friends who were so close. They'd be having fun, going places, and laughing, and not having to hide or be afraid or have a baby and live in a dumb old room all alone.

She sat up and wiped away her tears. She didn't care. She was going to call her friends and have some company. She had a right to live just like anybody else, didn't she? She couldn't stay cooped up in the room for weeks and months and years, maybe, hoping someday Link would get a transfer. He'd never get one. He never would ask. What was the matter with him anyway? Where was his spunk? Why didn't he tell that dumb gang off just like he used to tell off everybody else, even the police? Even . . . almost . . . her father. If he'd just do that, and be his old self, she would go to Chicago or anywhere with him, and be happy, and they could go out and have friends and have fun.

Not have to sit alone all the time with nothing to do except be afraid her father might catch them. If Link was scared, she wasn't. Her father wouldn't kill her, even if he found her. And she was going to see her friends. She had that right. She'd dress and go to her friends, no matter what anybody said. Link or anybody.

She heard someone coming slowly and softly up the stairs and made a grimace of displeasure. Link said he'd ask the landlady to look in on her. Did it have to be now? At the light knock she called, "Come in, Mrs. Stendell," and pretended to be greatly engrossed in the confession magazine she was reading, and wanted to keep reading. Link sometimes brought home a comic book, most of the time all he read were the car magazines; but she liked good love stories. More than once the idea had come to her that her own story was as romantic and sad as any she'd read. And someday she was going to write it down and send it in. They paid a lot of money for true stories of love.

"May I come in?"

She raised her eyes slowly, her heart seeming to float toward her throat. Fred Simmons stood in the doorway, dressed in beautiful clothes, looking like a movie star.

"Link isn't home," Darlene said, drawing her robe a little more closely around her. "He'll be back soon, if you want to come back later."

He stood in the doorway, his eyes taking in the picture of her. The way she was posed in that chair. Like a kitten. A soft, lovely, apprehensive kitten. Yes, it was going to work out fine. He knew that already.

"He asked me to stop by and look in," Fred said. "He was worried about you."

Link asked Fred to see her. That was a laugh. She knew how he felt about Fred. And what he'd done to Fred that night. On her account, she was sure. Although Link never would tell her what it was all about. But she knew one thing. Link would be wild if he caught Fred with her. She tried to do her part with dignity, as she had read about others.

"I'd rather you came back later, Mr. Simmons. I know Link will be angry if you are here when he returns." It was like a very little girl reciting a phrase she didn't understand.

"No, really," Fred said. "He asked me just before he left. He said you were sick, and would I come by and see you."

He sounded very convincing. He stepped one pace into the room. "I offered to get the car myself when I heard you were ill, but Link insisted on going himself."

Darlene turned her head to one side, puzzled by Fred's self-assurance, by what he was saying. "Link said you couldn't."

"I wanted to," Fred said. "Believe me. But he wanted to go. If you don't believe me, ask Link. I'll tell you exactly where and when I made the offer and he turned it down."

"I don't understand," Darlene said uncertainly. "Why would he want to go, when I . . ."

"There's probably some perfectly innocent explanation," Fred said. "Of course, he went there for quite a while before you started going with him, and there are some pretty girls across the river. And well . . . let's face it. If a wife is sick and another girl who likes you isn't . . . He wouldn't be the first."

"Not Link!" Darlene cried vehemently. "If he knew what you were saying . . ."

Knowing Link wouldn't be back, Fred made every move he wished in the game. "Ask him," he said. "Ask him when he comes home tonight. About Dotty . . ."

"Dotty?" Her face puckered slightly, and her lower lip trembled. Tears. He had wanted tears. How easy they were to bring.

"Well, look," Fred said, "I didn't come here to make you sad. I came to cheer you up."

"Go away," Darlene said. "Go away."

"I'm probably all wrong about Link," Fred said. "Let's not condemn him without a fair trial. That's not fair to Link."

"I know he loves me," she said, making a defiant face. "I know it."

"Of course he does. How could anyone help it? You'd be easy to love." He smiled, the friend-of-the-family type smile that was made by saying cheese silently. "Won't you let me sit with you a moment? I won't stay long."

"I don't know . . ."

"It's all right," he said, coming into the room, "Link and I have had our troubles, but we're all friends now. And I was best man at the wedding. Remember?" He tried to hold her gaze to make her remember the kiss he had given her.

"I'll tell you what we ought to do," Fred said confidentially, giving Darlene his most winning, youthful smile. "We ought to give old Link a dose of his own medicine. It would serve him right."

Darlene looked at him suspiciously. He looked like a movie star in his expensive clothes, and it was partly his good looks that frightened her. A small part of her wanted to be friendly with him, and she knew she shouldn't even want that.

He saw the appraisal in her eyes and exulted. She was falling for his charm, and his line.

"That's just what we ought to do," Fred said. "Give Link a dose of his own medicine. It's a shame the way he's kept you cooped up in this little room all the time. And running off to see that other girl the minute you didn't feel well. That's not fair at all, and we ought to make him pay for it."

She looked down, hesitant and doubtful. Not wanting to believe what he was saying, yet afraid he told the truth. She didn't know . . . didn't know . . . It was sinful to enjoy a strange man's compliments, but how could a girl help it?

"What you and I ought to do," Fred said, "is go out and have a real good time of our own. You put on some pretty clothes, and I'll take you to places Link never even heard of. I've got a new Caddy downstairs, and we can really have ourselves a ball. There's a place down by the airport where we can have a ten-dollar dinner, and dance, and have a few drinks . . . You deserve to have a little fun, Darlene. A pretty girl like you. You deserve some good times."

She couldn't help listening and agreeing. She was pretty, and she did deserve some good times before she got old. Before she had the baby to tie her down forever. And if what Fred said about Link was true, it would serve him right. She didn't have to do anything wrong, just because she went out with Fred. He was a gentleman.

"I don't have anything I could wear to a fancy place, even if I would go," she said. "And I'm not going. It wouldn't be right."

"We wouldn't have to go out if you didn't want to," Fred said. "We could teach Link his lesson right here. I've got some liquor in the car, and I could bring it up. We could have our little party right here, just the way you are."

"Oh no," Darlene said. "I'm not about to have any party like that with you or anybody else." She wasn't that dumb.

For a minute his line had sounded pretty convincing, but she saw through it all now. He approached her, his arms extended. "Aw, come on, Darlene," he said. "Don't be like that. You're so pretty . . . You just don't know . . ."

She jumped out of her chair and moved away from him. "You'd better get out of here," she said harshly. "I think you've been here too long."

"Listen, honey," he said, coming after her. "Be reasonable. I think about you all the time. I've just got to put my arms around you . . ."

"Get out of here!" she yelled. "Go away!"

He laughed and put his fingers to his lips. "Not so loud, Darlene. You wouldn't want strangers to come in and find us alone. You in a nightgown. Link wouldn't like that, would he? We might even be arrested together and have our picture in the paper. That wouldn't be nice, would it?"

She retreated behind the big chair, facing him like an animal at bay. She was frightened and afraid to show it. She tried to frighten him. "You better get out of here," she said, her voice harsh and strident. "I'll get Link after you. You'll need more than new teeth the next time my husband gets ahold of you."

Fred smiled, showing the teeth, toying with the scarf at his neck. "Maybe you ought to know," Fred said. "There won't be a next time. I'm a pretty important guy in the organization, and all I have to do is . . ." He snapped his fingers. "There goes Link."

"You wouldn't."

"Not if you will. Fair exchange? Your husband's life for something you'll enjoy? You can't go to the police, if that's what you're thinking. We're a big outfit. We'd have to get rid of you both. It's been done, you know."

He said it all with a smile, looking at her, sighing, trying to flirt.

She stood behind the chair rigid as a statue. There was nothing she could say, no way to escape. She was trapped and Link was trapped. People could do anything to them now, and there was no place to turn for help. No way to say no. This wasn't like being chased by a boy who liked you or just wanted you. It wasn't a game she could end with a slap. It wasn't anything real. It was like the nightmare of being on a track with the train coming and not being able to run.

She was trying painfully to understand how it could happen. All she and Link wanted was to be married and left alone. They didn't ask to be chased out of Dellville. Link had tried to find a decent job. And even if he was driving some stolen cars to Iowa, why did it have to lead to this? What did that have to do with this?

No place to turn. Not for Link or for her. It was as Link had said. The only place they could go was in, deeper and deeper. This was deep. Why? Why? Just because Link drove their cars, why did she have to submit to this man? Why could somebody threaten to kill you, and you couldn't ask anybody to help you? Just because he drove their cars. And after this man? Who else would want her, and she'd have to say yes, or Link would be hurt? Whoever did, whoever it was, could have her. Then Link wouldn't want her. Then what would she be? Then what could she do? What would become of her and Link and the poor baby? Oh God, how had it happened . . . ?

She made a last, quietly desperate, naive plea. "Please," she begged, her eyes fixed vacantly on his scarf. "I'm going to have a baby."

As he came up quietly behind her and bent forward, his hands on her shoulders, his lips touching the back of her neck, Darlene felt herself beginning to shudder . . .

CHAPTER 22

VIRGIL KERN SAT AT THE POLICE DESK in city hall, one leg hooked over the arm of his chair, his cap on the back of his head. A sleeping drunk snored in the one cell at the rear, the city clock ticked off the night with the loud, monotonous tolling of its kind. Every sixty seconds, the large minute hand leaped a fraction of an inch as though the victim of a metronomic spasm.

The town outside was quiet, but it caused Virgil no joy. What did it matter now how peaceful the town was? How law-abiding? What difference did it make now? The one thing he'd come for, and given up everything in Des Moines for, that one thing had blown up in his face. That's the way life was. A man gave up everything to protect his daughter from evil, and when he got through running from it, there evil was, ahead of him, and waiting for him. Wonder where she was, and if she was all right. If she had to run off after that punk, the least she could do would be to write and let them know she was alive. But if she had sense enough to write, to think about her folks' feelings, she'd have had sense enough to stay home.

A man was a fool to try to do anything for anybody else. After all he'd given up for her, though. And she didn't appreciate it. Did everything she could to wreck what he was trying to do for her own good. And it was a wreck, all right. She was gone, God knew where, doing God knew what to keep alive. If she even was alive... And Agnes blaming him because Darlene had run off. And he was the only one who had tried to keep her away from that punk. That was Agnes, though. Let him give up a good job and try to find a decent home and decent friends for the kids. Nobody lifted a finger to help. Let things go wrong because other people wouldn't help, and he got the blame. If that wasn't something!

He'd done all he could to protect the girl, and she was gone. She was gone, and here he was, stuck in Dellville for no good reason. Sitting alone in city hall, listening to a drunk snore. If that wasn't a life for a real policeman. Maybe he could get back on the Des Moines force again. Ride the night patrol with Ernie Cross again. Have somebody to talk to and not be so lonely all the time. Have work to do. Real work.

He recognized a heavy footstep outside and curled his lip in disgust. That Dutchman. He shook his head. He'd sure got himself into a fine spot all right. All on account of Darlene, and for nothing.

Arnie came in and sat down, folding his hands across his stomach. "Quiet tonight, eh, Virgil?"

"Always is."

"It's nice that way."

Virgil rolled a cigarette, listening to the clock and to Arnie's heavy breathing.

"I sit a while with you," Arnie said. "Get used to the night hours. Soon you'll take the days, and I'll take the nights. A married man should be home at night." Arnie chuckled. "Two policemen in Dellville. One for the day and one for the night. We're a regular city. Don't see how I handled the whole job alone for over thirty years. Ya, I remember when I started here. I had a chest then. It was different in those days. Horses you saw, beautiful horses. There was a pair of matched bays, you never saw anything prettier . . ."

Arnie rambled on, his eyes half-closed behind his rimless spectacles, his head nodding gently. Kern smoked his cigarette and tried to close his ears to the old man's muttering. What he wouldn't give to be on duty with Ernie again, where there was some law to enforce, and somebody to enforce it against.

The police radio cackled like a sick chicken. The voice that broke into the static was the usual toneless, distinct official voice. "Attention all state and local officers . . . Lincoln Emory Aller, repeat, Lincoln Emory Aller . . ."

Kern's foot slid off the chair. He reached for a pencil. Arnie's lids raised slowly. The blue eyes stared sadly at the radio.

"Lincoln Emory Aller. Age nineteen. Five feet eleven. One hundred and sixty-five pounds. Black hair, black eyes. Last known address seven four one Eighth Street, Des Moines, Iowa. Wanted in connection with car theft and assault against a police officer. Warning. This man may be armed and dangerous. Approach with caution. Repeat . . ."

Kern got to his feet. "Take over, Arnie."

"The boy won't come here, Virgil."

"I ain't looking for him." Kern folded the paper he had written on. "That's his address and it'll be her address too. Des Moines. I never thought they'd stay in the state. Or I'd have looked there. I'm goin' there now. Maybe I'll find her. And when I do . . . !"

"Don't be too angry, Virgil," Arnie said. "Maybe it ain't so bad as you think."

"Ain't so bad! Livin' with that punk? Afraid to write home! Sounds like a respectable married couple, don't it? Well, she's lived with the punk all she's goin' to, I'll tell you. I just hope he's there, that's all. That's all I hope."

"Ach, Virgil..." Arnie tried to calm his fellow officer. "Give her a chance to tell her side. Don't make it worse. If the boy is there, he'll be scared. Be now a friend, eh? It's too late to undo what is, Virgil. Maybe you can help now..."

"I'll help, all right," Virgil said, licking his lips. "It's easy for you to talk, Arnie. You ain't got a daughter that's been ruint and disgraced in front of the whole world, and people laughin' behind your back about it. You don't know what it is to see your only girl dragged through the mud by a no-good lousy... I'll help 'em, all right. Help him into the next world if I get my hands on him. And help her to the worst whipping she ever had. She won't be able to set fer a week, I'm telling you."

Kern's Lincoln was parked outside. He slid behind the wheel, kicked the engine into life, shifted into low, and floored the gas pedal. The big car screeched over the cobbled square, motor snarling as he raced toward the highway. Arnie went to the door and listened as the engine sounds faded. It didn't matter who it was, that sound was always the same. A speeding cop sounded just like a speeding hot rodder. "I know it ain't right," Arnie said apologetically to the clock. "It ain't right to say this, but I hope he don't catch the boy. Not right away. Not tonight."

— — —

Darlene stood behind the big chair, gripping its back with her hands, her head bowed. Fred stood beside her, gently stroking her hair. He wondered if he ought to leave his dentures in or put them in a safe place. He hadn't had them too long, and he wasn't sure he could control them in every situation. It would be embarrassing if they got loose or fell out at a critical moment. A mishap like that could ruin everything. Yet, he looked so awful with his teeth out. The problems a man had... He clicked his teeth together and looked at Darlene, admiring her. He hummed faintly, thinking of Link. That miserable animal...

He continued to stroke Darlene's hair gently. She wasn't

shuddering anymore. He smiled. The police would catch Link, and Link would be a sullen, uncooperative prisoner. He would be defiant and refuse to talk . . .

Fred looked at his watch. Perhaps they already had Link and were beginning to work on him. If everything went well, Darlene would feel friendlier about the time the police were landing on Link with both feet. It was something to marvel at. The finest gag he'd ever pulled on anyone. The beautiful timing, the wonderful feeling of revenge that would be his, knowing how Link would suffer. No hard feelings, eh, Link? Fred laughed silently and stroked Darlene's hair.

His hand stopped moving, resting against the back of her head. His muscles tightened.

Someone coming?

A soft step outside the door. It couldn't be Link. It couldn't be. A stair creaked.

There was a light knock at the door. Darlene quivered. "Whoever it is, get rid of them," Fred whispered, digging his fingers in her arm.

There was another knock. Louder, more insistent. It was followed by a harsh voice. "You in there, Darlene?"

She cried out before she was aware of what she did. "Daddy!" she screamed. "Help me! Daddy!"

The door cracked under Kern's shoulder. Kern charged into the room with his blackjack ready for action. This time he was going to kill Link.

Kern stopped with his arm back when he saw Fred. He had been so sure of his target, he was temporarily shocked into immobility.

"Help me," Darlene cried. "He . . . he . . ."

"Don't you try to get tough, fella," Fred stuttered. "You . . ." He saw Kern's police uniform and his mouth sagged open. That was when Virgil swung the sap, and the target he aimed at and hit was the two-hundred-dollar dentures.

It wasn't noisy. Kern did everything Link had done to Fred, then added extra punishment of his own. He knew how to beat a man, and he gave his finest performance.

Darlene watched from the bed, her hands clenched into fists, hitting the air whenever her father hit Fred as though to add her strength to the blows.

Kern finished and threw Fred aside. Then he turned on Darlene. "Just what I expected," he said thickly. He strode toward her, raising his hand to strike.

"No, Daddy!" she cried, trying to get out of his way. "I married Link. We're married. This . . . this . . ."

"What about this?" He stayed his hand.

"He's Link's boss," Darlene sobbed. "Link had to go away on his job, and he came here, and . . . and . . ."

"I know the story," Kern grunted savagely. "Married!"

"I can show you the license, Daddy. Honest! And Link has a job now."

Kern snorted. "He's got a job all right. Yeah! Come on. Get your things. You're going home with me."

She shook her head. "No, Daddy. We're married. We're going to have a baby. I have to stay with him, Daddy."

"You can't," Kern said roughly. "There's no him to stay with anymore. Ain't you heard?"

She shook her head wordlessly. Dead? An accident! He was hurrying to get home. An accident!

"He stole a car," Kern said. "Beat the policeman who stopped him so's the fella will maybe die, and run off in another stolen car. It was all on the police radio."

He waited while she sat hunched over and wept bitterly into her hands.

"You wouldn't listen to your daddy," he said roughly. "You knew better than your daddy, didn't you? You thought your dad was an old fool and didn't know nothin'. I could have told you somethin' like this would happen. Ain't I seen enough of it? Wasn't I always scared it would happen to you? And it happened, didn't it? Just like I said it would."

"What am I going to do?" she whispered through her fingers.

"Come home where you belong," Kern said. "Come on, Sis, I'll help you pack your things. And have this fella locked up."

She looked at Fred, and remembered that he was part of a gang, and if anything happened to him because of her, they might hurt Link. She shook her head. "Let's forget everything," she said. "Let's just forget. I've had enough trouble already. Why did it have to happen, Daddy? Why do I have to have so much trouble?"

"You're startin' to talk like your ma," he said gruffly. "Come on, now, let's get your things. I'll send the ambulance over to take care of that thing." Virgil saw the intact lower denture on the floor. He stamped on it with his heel until it was smashed.

"Young fella like that wearing false teeth," Virgil grunted. "You'd think he wouldn't need them at his age."

"He did have nice teeth until a while ago," Darlene said. "Link knocked the real ones out."

"I ain't surprised," Virgil said, helping Darlene get her clothes. "It's just what you'd expect from a hoodlum."

They drove back to Dellville in the Lincoln, at exactly sixty miles an hour. They were silent, thinking. She was back, Virgil thought. The way he'd always feared. No husband, a baby for him and Agnes to take care of. In disgrace. After all he'd done to keep her decent. She'd sure gone whole hog.

And yet, now that it was so, now that his nightmare about her had come true, it wasn't so bad. She was still Darlene, his daughter. She still looked the same on the outside. It was the thinking, the crazy pictures that had come into his mind, that had been the worst thing. And now . . . well, it had happened, and there wasn't anything you could do to stop it. You had to be practical and make the best of the things life handed you.

"Oh, Link . . . Link . . ." Darlene cried to herself. "What's going to happen to you? How will I know where you are? How will you know where I am? If I could just let you know I love you . . ." She sniffled.

"Cryin' won't help," Kern said, not unkindly.

"We were gonna buy a house someday," Darlene said forlornly.

"And buy a big couch with feather pillows that costs six hundred dollars. We had it all picked out."

"Don't start bawlin' about that," Virgil said. "Six hundred dollars for a couch. It'll be just a while before it gets snapped up at that price. Probably still be in the store a hundred years from now when Link gets out of jail."

There was a tearful reunion with her mother, and Virgil broke the news about the baby. That sent Agnes off in a long scold as to why Darlene couldn't have a baby because they didn't have any room for a baby, and she was too old to start that kind of work again, staying up all night walking the floor, and Darlene must be hungry.

"I don't want anything," Darlene said. "I'm not hungry."

"You better eat something," her mother said severely. "I ain't thinking about you, Darlene. I'm thinking about the other one you're eatin' for. You're drinking a big glass of warm milk!"

It was very late when Darlene went to bed, and she was exhausted. She tried to worry about Link, but the minute she lay down and put her arm around her old teddy bear, she was asleep.

— — —

"Chicago," Charley said, trying not to sing it, "is a wonderful town. How do you like it so far?"

They were driving along Lake Michigan's shore, the dark water on one side, the massed lights of the city on the other. Lights of a million cars swarmed around them like the bugs that used to swarm over the yard light Link had on the garage.

"Looks fine," he said. How could he let Darlene know? She was alone, sick, waiting. She would worry if he didn't come back. She'd hear and worry. How could he reach her? Let her know? She was alone and sick.

"How can I get in touch with my wife to tell her I'm all right?" Link asked Charley.

"Who? Your what?"

"My wife. She'll worry."

"Mrs. Aller?"

"Yeah."

"What's Mrs. Aller to you, Mr. Richard Monroe? You don't look like a guy who'd play with another man's wife. You stay away from that Mrs. Aller."

"I'm not kidding, Charley."

"Neither am I. You're hot, kid. You heard the news. That cop might die. They'll be looking for you. Watching Mrs. Aller to see if her husband tries to get in touch. So, Mr. Richard Monroe, he never heard of Mrs. Aller. You see, kid, I didn't explain about that new name routine. When you cut off the old name, you cut off everything that goes with it. I mean, if you want us to keep you on the outside."

And looking at Chicago, so different from anything he had ever seen before, so far from Iowa, and Des Moines, and Dellville, so far from Darlene, he did feel that he might be Richard Monroe, and that Link Aller and everything that was Aller's were part of something he'd heard about or knew about and couldn't remember too well.

He lit a cigarette and stuck it in the corner of his mouth.

He'd come a long, long way from Dellville. Not too many weeks ago he'd been driving a grocery truck, nothing but a hick. Now he was Richard . . . Ricky . . . Dick Monroe, a Chicago boy. Just like that. One of the Chicago boys.

CHAPTER 23

BEING ONE OF THE CHICAGO BOYS wasn't what Link had expected. He didn't know quite what it would be like, but from what he had heard, he got the idea that the organization was a band of modern jolly robbers who all lived together in a great big castle with their chief at the head table at dinner. And from which stronghold they sallied forth to rob the rich of their gold and cars.

Nobody had said it was like that, and he was pretty sure it was a silly picture he had, but the phrase "Chicago boys" had always made him imagine a bunch of friends together, laughing and talking, and plotting perfect crimes. Like he and the guys used to be in Dellville, when they'd gather in the drugstore and plan which town to bedevil in their cars.

Anyway, he thought there would be something. But there wasn't anything.

Charley had driven him to a place on the South Side that was about the crummiest, rattiest place Link had ever seen. A broken-down, dirty place with the brick crumbling, and some windows broken out, and garbage and stink all over the place. They went through a dark hall with a half-exposed toilet at the end, the whole place reeking like an outhouse, and up some stairs. Upstairs there was a small room with an old sagging iron bed in it and no linen. But there were some newspapers spread on the cotton mattress. There was a broken-down dresser in the room, and a straight chair. The walls were cracked and dirty, and a bare light bulb, very tiny, hung from the ceiling.

"Here you are, Dicky," Charley said. "Home sweet home. Palmer House Annex. The maids haven't had a chance to clean the room since the last guest was eaten by rats, but they'll be here to tidy up in about . . ." he looked at his watch, "five minutes after the Second Coming. Like it?"

Link's stricken look was answer enough.

"What'd you expect?" Charley said. "A brass band and a suite overlooking the lake? Just because you cold-cocked a cop? It's better than prison, my boy, much better. Not as clean, I will admit, and there ain't a library or a sports program, or school classes or a bunch of fellow workers to chin with, but here, my boy, here you are your own boss. And you don't have to get up until you feel like it."

"What do I do?" Link asked, an awful sensation, like homesickness, digging at his stomach.

"Nothing," Charley said. "That's the simple beauty of it. You lie down and get comfortable, and when that gets uncomfortable, you change to another position."

"For how long?"

"When you're tired of laying down, you're free to get up and sit in the chair."

"No," Link said, trying to fight the feeling of oppression that was taking the heart out of him. "I mean how . . . how . . ."

"I know what you mean," Charley said, looking under the newspapers on the bed for bugs. "You got to check on these innkeepers," he muttered. "I bet these newspapers haven't been changed for two weeks. I know what you mean, Dicky." He stood close to Link, the cigar in his mouth, tapping Link's chest with his finger. "You've been a bad boy, Dicky, and we have to keep a lid on you, or you'll be collected. Right now, the hunt is on, and your face has to stay out of the street."

"But how long?"

"Maybe forever," Charley said. "Depends. If the cop lives, the heat will cool after a while. You'll just be a name on the list. If the cop dies . . ." Charley rolled his eyes toward the ceiling.

"How about eating? I get to eat, don't I?"

"Downstairs. You eat with the guy and his wife who own the place. They'll call you or feed you up here."

"And I just stay in this room?"

"You can use the plumbing in the hall."

"Thanks."

"Unless you hear anybody else out there. Then you wait until you're alone. You don't know who's apt to peek in here, you know?"

"I thought you Chicago boys did better than this," Link said.

"We do," Charley said, poking Link's chest again. "But we ain't hot. What'd you think we had here? A special hotel for fugitives, with no cops allowed?"

"After all the big talk . . ."

"You think this is free?" Charley asked sharply. "Who do you think is paying the tab for you? You got the dough? You think people hide killers for free?"

"Okay," Link said. "I'll sit it out."

"I'll bring you a deck of cards," Charley said. "And something to read."

"Some hop-up magazines would be good," Link said. "And some comics, for fun reading. If you don't mind."

"Anything for a buddy," Charley said.

"Charley," Link said, "how am I gonna get word to my wife about this? She's sick, and she won't know . . ."

"The papers, the radio," Charley said. "They'll tell her."

"They won't tell her that I'm all right," Link said. "If she could just know that . . ."

Charley shook his head. "You're not all right. You're nothing right now, Dicky. You forget you ever had a wife. Especially if you're stuck with a murder rap."

"But Charley . . ."

"Who do you think you are?" Charley demanded. "Link Aller with a new name? You think that's what an alias is? Kid, you're Dick Monroe—until you get caught. Dick Monroe. Don't you understand?"

"Sure," Link said. "I understand." But he didn't, he found out later. He didn't understand at all.

"Of course," Charley said, licking his cigar, "I could have Fred get in touch . . ."

"Never mind," Link said slowly, hating Charley's seemingly innocent smile. "Just never you mind."

"Okay, Dicky," Charley said. "Have it your way. And start farming while you're here." Charley laughed at Link's expression and touched his upper lip. "Grow it there, boy."

"I hate mustaches," Link said. "The . . ."

"Grow it." It was an order.

Link was about to tell Charley that there were some things nobody could decide for him, but he looked at where he was and knew why, and he repressed his anger and said, "Okay, Charley," in a beaten, tired voice.

Charley left and Link lay down on the bed. He felt as though he had been buried alive in a filthy coffin. The dim bulb in the socket added to the gloom. Even a brighter light would help. This way, in a dirty half-light, it wasn't like being alive.

He listened to the strange city noises, the faraway grind of streetcars, horns of cars, the echoes of life. He wanted to get up and run out of this miserable, stinking room and go where there were people. Not to be so alone. All alone.

It was almost dawn. Darlene. Was she staying up all night, waiting for him? How would she know what to do? Darlene . . . He turned over and lay face down, and cried, but even in his grief he was choked by the stink of the mattress and had to turn on his back to cry and hit the mattress with his fists.

Finally, he wiped the tears from his cheeks, feeling the stubble of whisker on his face. Hard guy. One of the Chicago boys. Cold-cocked big cops. Crying. Link Aller crying. No. Dick Monroe crying.

What a stinking hole.

In the morning, Charley came back with some magazines and another man. A thin, pale man in a black summer suit. With an impatient, mean look. With bad teeth, circles under his eyes, and a rasping voice.

There were no introductions. The new man looked at Link, sucked his teeth, and began to bawl Link out like a little kid.

"So, what's the matter with you? You think you were hired to fight cops?"

"He caught me," Link said. "I had to do something."

"So, he caught you," the thin man rasped. "So, you get six months or a year in the pen . . . Maybe two years. Is that gonna kill you? You go in for a while and it's over."

"I had to get away if I could," Link said. Charley raised his eyebrows.

"Why? You call this getting away? Instead of a car rap that leaves everything quiet, you go stir up the whole country. You go for a murder rap. And some bridge guard remembers you, and the cops are crawling all over the place looking for a hot car ring." He turned to Charley. "That's the trouble with these punk kids. No brains. They think with their hands and feet. That's why they're no good to us. You hear what this dumb farmer tells me? He's gotta get away from the cop. And look what he does to us. Makes us close down all over the place. The money we lose! And then we have to take care of him, so he stays out of trouble."

Link was red with shame and embarrassment. To be talked about like this. As though he wasn't even in the same room.

"You stay in this room," the thin man said, showing his bad teeth. "And don't stick your nose outside for nothing. Not until you're told to. You go outside this room, and you're in real trouble. From us. You understand?"

"Yes, sir," Link said. It was like getting bawled out by a cop.

"And if they do catch up with you, you don't know anything. You were on your own. You know that?"

"Yes, sir," Link said. "That's what I'll tell them. That I just did it on my own."

The thin man looked at him contemptuously. "It won't be so easy when they start in with the rubber hoses. But hoses are easy compared to what we do if you squeal. You understand?"

"Yes, sir." The life had gone out of Link's voice.

"You remember that." The thin man turned to Charley.

"The trouble we get in with these punks. They ain't worth it. We take a couple of years to build up some good outlets, and in one night this punk and that Fred, they louse it all up. And want a thank you for it."

"Fred?" Link said. "Fred Simmons?"

"The guy that hired you," the thin man said. "You two, you're a pair all right. You belong in the same box. You slug a cop. Fred, he gets slugged by a cop."

Link's mouth twitched, but he kept back the smile. "What happened?"

"Nothing," Charley said. "You know Fred."

"Nothing," the thin man said sarcastically. "Less than nothing. He had to go for some girl. So, he's with this girl, and her father kicks down the door, and it turns out he's a cop. It was our night for cops in Iowa last night. Fred's in the hospital, you we have to hide, and the whole route there is shot. I tell you, Charley, these young punks ain't worth the trouble they make. Fred, I have to bring back to Chicago when he gets out of the hospital, and this one . . ." The thin man looked at Link with a half-humorous, half-threatening expression. This one we might have to drop in the lake if that cop dies. Come on, Charley. We've got business."

Link didn't hear any of the final talk. He went back to the bed and sat down, his hands hanging, his shoulders bent. There couldn't be two!

Maybe when he'd been hitting that cop, fighting for his life, maybe then was when it happened. Fred with Darlene. Her playing sick. It all fit. Fred with Darlene. While he was fighting that cop.

Yeah . . . that was the first time there hadn't been an Iowa kit on the car. And an old car. And all of a sudden, a cop on his tail. Like the cop knew. Sure, the cop knew. Sure!

Link made sounds like a dog whining as it all fit into place. The finger on him, to get him out of the way, and then Fred and Darlene . . . while he was fighting for his life. Until Kern had found them. That was funny. Kern had found them and put Fred in the hospital. And Darlene . . . ?

He sat on his bed and stared at the walls that were his home and his prison. God alone knew what had been going on. With Fred . . . Darlene . . . If he could only get out and find out for sure. If he could only get out and do something!

He jumped to his feet and paced back and forth, raging, tormented, wanting to howl out loud with grief and anger and doubt. Wanting to know! A man couldn't not know about a thing like that. Where were they? What had happened? God, what had happened to her? To him? To the whole world?

"Hey, you up there, mister!" The woman from downstairs was calling through the hallway.

"Ahh?" He half-roared it.

"Stop the tramping, please. You're shaking down the plaster from the ceiling."

Her door closed.

He took a step, halted, and fell on the bed. He couldn't even walk! Couldn't walk, couldn't shout, couldn't do anything. Just stay in the filthy, stinking room, sleeping on dirty newspapers, not knowing anything. Walled in. Trapped. Buried. Alone . . . Alone . . . "Oh dear God," Link implored, lifting his dirty gaunt face toward the cracked ceiling. "Help me, help me . . . I'll go nuts . . . crazy nuts in here. What's happened, God? How? How did I get here? I want to know about her, God . . . I've gotta know. Please God, help me and I'll be good. Just make it not be her and I'll go to church again like my mother said I should. Honest, God! Honest to God . . ."

He let out a long sigh that was sadder than tears and more desperate and despairing than prayer. His soul ran the white flag of surrender, whipping up his throat in that last, long sound.

He picked up a copy of *Hot Rod* magazine and tried to read, but he couldn't. The walls had pressed in so close that they squeezed his eyes together, and he couldn't see.

He lay back on the bed and smoked, staring at the ceiling. But there was no lying still. His body hummed, vibrated, seemed to be writhing under the skin. Something had to happen soon. He couldn't stay like this. They were changing the world on him beyond that window shade. They were changing the world all around on him, and he wasn't allowed to see it.

His foot was tapping rapidly against the iron end of the bed, beyond his control. It began tapping faster and faster. "Dah-dah . . . dah-dah . . ." he hummed shrilly, trying to keep up with the foot. "Dah-dah, dah-dah-dah-dah-dah . . ." He began to sing jerkily, his body rigid. "Dah-dah, dah-dah, I'm a Chicago boy, a big Chi-ca-go boy . . . dah-dah, dah-dah . . . Darlene!"

He stuffed his hand in his mouth to hold back his growls and cries. Darlene and Fred . . . Fred and Darlene . . . while dah-dah he dah-dah was a dah-dah, dah- dah, a fighting for his dah, dah, dahhhhhhhhh . . .

"Let me out of here!"

Dah-dah, dah-dah . . . . . . big Chi-ca-gooo boy.

"OUT! LET ME OUT OF HERRRRRREE!"

"Hey, you up there, mister, you want to wake up the dead?"

"It's like this, God," Link said rapidly, in a low, confidential voice. "My name is Lincoln Emory Aller, only You probably heard me called Link. I used to live in Dellville, Iowa. I think You know my mother there. I had a race with a kid, and he was killed, God. His name was Ricky Madison, and I think he's up there with You now, and his girl, Sharon Bruce. It wasn't my fault, God, was it, Ricky? I mean I didn't push you the way they said I did. You were way ahead of me, Ricky. Honest. You really had some rod there. I was only kidding you when I said it was no good. God knows I was only kidding to get you mad, Ricky, wasn't I, God? But they all said it was my fault and none of the guys were good friends after that and then they all went away and nobody but kids left. And they didn't ask me to join the Timing Association, and I made believe I didn't care, but I'm telling You the truth, God, I wanted to belong to it, but I didn't want to ask. I didn't think I should ask, me being an older guy, and I guess that was the Devil making me proud, Lord, and I had to show them I didn't care. But I cared, Lord. I can't lie to You. And then I met this girl who came to town, God, and You would have liked her, Ricky and Sharon, and the four of us could have had a lot of fun, only her father beat me up, Lord, and drove me out of town and there wasn't any jobs, Lord. You know that, I guess. Maybe You planned it that way to make me suffer for my pride and sins, Lord, and I started diving hot—that's stolen—cars, Lord, and I guess You know what happened then, Lord, and how it was all framed against me by Fred . . . And, Lord . . . God

... You know we were married, me and Darlene, and she was going to have a baby, and then this thing happened, and Lord ... oh, Lord, I've been awful wicked, I know, and I deserve anything You want to do to me. And I guess You know all about everything that happens, Ma says, and will happen, and, Lord, please Lord, I'll take anything else You want to send down to punish me and I won't care, but God, Lord in Your Heaven, don't let it be true about Darlene ... Please, Lord ... I'm begging You on my knees like I never begged for anything before, God. You know she's the only one on Your earth ever treated me nice and don't hurt me that way ... don't ... And if You had to Lord, and that was Your way, I'm sorry I asked You about it too late. And Lord ... please let that policeman live so I can make it up to him some day and look out after Darlene while I'm away from her and I know You didn't let it happen, that's what I'm thinking, Lord. I really don't believe You would. You wouldn't let any hurt come to Darlene because she's a real nice girl and I know it wasn't her, and Lord if You can, please send word to me which other policeman's daughter it was ... I'm sorry I bothered You so long, Lord. And I don t expect You to do anything for me, but I sure would appreciate it if You didn't let it happen, Lord. Thank you, Lord. And Amen."

Link sighed again, and lay back on the bed, tired, but feeling a sense of peace, a little sense of peace that bobbed uneasily on the tides of doubt and suspicion and despair that surged in his heart.

"And Lord," Link said quietly, humbly, with a willing meekness he had never felt before in his life. "If You're still listening, I want to tell You. If I ever get out of this and ... everything is all right, I want You to help me be a good Dellville boy, Lord. I don't want to be a Chicago boy."

CHAPTER 24

THREE WEEKS LATER, as Charley was leaving Link's room after a visit, he paused at the door. "By the way, kid..."

"Yeah?" Link was wearing a pair of blue gabardine trousers, low shoes, and a sport shirt. Charley had thought it would make him feel better to be dressed a little. He had spent a great deal of time watching the mustache grow, learning to trim it, and training it to lie right. He thought he looked good in a mustache. It made him look older, more worldly, and a little sinister.

"How would you like to leave this joint with me? Now."

Link looked imploringly at Charley. "Don't kid me that way, Charley."

"No, kid," Charley said grinning. "You're sprung. The cop's gonna live, and the heat has been turned off enough so you can show your face again if you're careful. You know, you're lucky. Most nineteen-year-old guys who try to grow a mustache look like high school boys with pussy willows pasted under their noses. But you've got that mean face, and all that black hair, and you could pass for thirty. It looks good on you, not like you went to the drugstore and bought a ready-made for a disguise."

Link was ready. All he had to take with him were some clippings he had saved about working on engines.

"It wasn't so bad was it, three little weeks?" Charley asked.

"The first couple of days were bad," Link said. "Then I got used to it. I sure caught up on my sleep. What's on the hook for me?"

"Now we're gonna make you a real Chicago boy."

Link nodded with grim satisfaction. "That's what I've been waiting for."

He had it all worked out. It hadn't been easy, but he'd done it. Link Aller was gone forever. Dick Monroe had taken his place. And Dick Monroe had never even heard of Dellville or a girl named Darlene Kern. Thinking about it that way, it was surprising how it didn't bother him anymore.

Being a Chicago boy now was a little different, but still short of Link's early ideas. He was put on the payroll, but he worked for his money in a garage. It was just like any other garage, and he didn't get into any mob activities or know which of the people he dealt with were also Chicago boys. He did his work, drew his pay, and was bawled out if his work was sloppy, and he was expected to put in his time just like any other job.

He moved to a better room, but it wasn't a gang stronghold of any kind. It was just a room in a rooming house. He was a mechanic. Other people who rented rooms there included a couple salesmen, a teacher, and a fellow who worked for a radio factory.

Charley had helped him pick out some "Chicago" clothes. "You're through wearing those cowboy duds," Charley said. "That's kid stuff. And country. We have to make you look as though you were born on the corner of State and Madison and was never any deeper in the country than Lincoln Park. Change your character to go with that tickler under your nose."

He made Link buy a double-breasted blue suit with a white pin stripe, a hat, and shoes with pointy French toes that almost crippled Link before he got used to them. He got a supply of white shirts, and a few colored ones, with round collars, spread collars, and button-downs. And some ties.

When he was all dressed in his new clothes and saw himself in the mirror, complete with a tie clasp and a handkerchief in his breast pocket, with his mustache and bold eyes, he had to admit that he really looked like a smooth city operator. Like a real Chicago boy. Not much resemblance between this guy with the French cuffs and gold cuff links and the Link Aller who used to think he was dressed up if his overall pants were clean.

Darlene . . . The girls ought to see him now, Dick Monroe, the natty city fellow with the big city way.

"When you walk," Charley said critically, "keep your legs together and take shorter steps. You walk like you were trying to straddle a furrow behind a plow. Make small movements, be graceful, don't let policemen look you too long in the face, and one of these days you'll make the FBI most wanted list for something dignified, like stealing the gold off the capitol dome."

The past was almost forgotten. Almost.

Charley came into the room one day and saw a copy of the *Des Moines Sunday Register* on the floor. "Where'd you get this?" he asked Link.

Link was reading the funnies. "I buy it every week. Just to see what's doing in the old home state."

"You'll be doing," Charley said. "Doing time. Don't you know the cops have all these out-of-town newspaper places watched? If I had a dollar for every guy who's put the finger on himself because he had to read his hometown paper, I could go into publishing. I'd like to be a publisher. A small sheet, about the size and color of a ten-dollar bill, with the same news and pictures on it."

"You think they're still looking for me?" Link asked.

"Still, now, and always. They might not look hard, but they're looking. It's a habit with the cops."

"Maybe I could get some kid to buy it for me."

"Yeah. And he has a policeman to help him deliver it. This might be a shock to you Iowa people, but you would be surprised how few Chicago kids read the Des Moines paper."

"I could get it in the mail. A lot of people do that."

"Sure. Hand out your name, address, and the hours you can be found at home."

"What am I supposed to do, then?"

"Read the Chicago papers," Charley said.

Link had taken a lot of orders from Charley and the organization, and he didn't like it. Once, when he was a raw country kid, he would have shot off his mouth about it and got in trouble. But he had learned, being in Chicago, how to seem to take their orders without letting them know what was in his head.

They could call him Dicky Monroe all they wanted to. He knew he was still Link Aller, and some day he would go back to being Link Aller. Someday he'd shake them all and go back to being himself—when the time was right.

Link smiled at Charley. Charley thought he knew everything, but there was a lot Charley didn't know and wouldn't find out until it was too late to stand in the way. Charley didn't know how many times Link had gone down to the post office and bought a card to write to Darlene—or how many cards had been torn up without his mailing them.

Now why hadn't he mailed Darlene a card? It had been simple enough in Des Moines, and there wasn't any reason the same trick couldn't be turned from Chicago. All it took was two cents to let her know he was alive and well and to get word from her in return. And after that, he wouldn't be alone any more, and he'd have somebody close to remind him who he really was.

Time and again, when feeling a wild loneliness, he had bought a card and written to her. But every time, after the card was written, he had torn it up. There were so many things that seemed to hold him back. He got afraid of what would happen if the Chicago people didn't want Darlene around and got mad at him for writing. And he remembered how weak and helpless she had been the last time he'd seen her. He was afraid she might still be that way, and he would have to take care of her instead of her taking care of him. And somehow, despite all the abuse he had to take, the gang gave him something he needed. It took care of him. It saw that he had work, and money, and a place to stay, and had plans for his future. In a lot of ways, it was very comfortable to go along with what he was told to do and let somebody else do the thinking for him. It was like being taken care of. Almost like the way Darlene had looked after him. Of course, he was going to break away some day, and go back to being his own self, with nobody to boss him, but he had to wait until the right time. That's one thing he had learned. How to be patient and wait until the right time. When that time came, he'd make his break, hunt up Darlene again, and, if everything was all right, take up with her and be himself again.

And it was because of his dream someday to regain the past that he fought Chicago in one strange way. No matter what else he did, he avoided getting a girl. As long as he could keep Darlene as the answer to his needs and loneliness, he felt he had a chance to break away from Chicago and return to his true self. He was afraid that if he let another woman take care of him, he would forget Darlene and want to be Dicky Monroe, and that would be the end of him. And then there were times when he thought of her and Fred and he hated her and never wanted to see her again—until he remembered how they had been in love, and he wanted her.

Someday he would get in touch with her and get everything straightened out. When it was time. That was the smart way. Just play along and wait for the right time.

Link was getting tired of having Charley run his life.

Charley told him where to live, and what to wear, and where he could go and where he couldn't and what name to use and how long his mustache ought to be. And now even what newspaper he had to read.

"Be a good boy," Charley said. "You be a good boy now, and it will pay off later. What the hell, Link. I don't want to be your nursemaid. But you flew off the handle with that cop in Davenport, and the boys want to take a longer look at you before they give you any work. Have to make sure you're steady. You know. You have to earn their faith and confidence."

"All right," Link said. "Have it your way."

Charley got up, stretched, and scratched his chin. "Old friend of yours is coming to town," he said lazily.

"Who?"

"Fred. He's healed up now, and we're bringing him back. He needs to live a more active life than he had in Des Moines."

"Fred, huh?"

"Get that look off your face," Charley warned genially. "You and him are gonna be great friends now."

"Look, Charley..."

"All that other stuff is dead, ain't it?"

"Sure."

"Sure?"

"Sure, I'm sure. I know how I feel, don't I?"

"Then what's with Fred?"

"You know, Charley."

"No," Charley said flatly. "I don't know. You tell me you're finished with that girl, and then you want to beat up on Fred on account of her. Come on Dicky, make up your mind. This I have to know now. Is that stuff over, or ain't it?"

"It's over." Link chewed at the corner of his mustache.

"Now, you're sure."

"Yeah."

"And hands off Fred?"

"If you say so."

"All right. It's a funny thing," Charley said, "but the middle boss—the one you saw—he's got an idea that you and Fred ought to make a good team for us. He figures that Fred was the one who found you, so you're a natural team."

"Wait a minute," Link protested. "I might not be able to do what I want to when I see him, but nobody can make me be his buddy. You couldn't ask that."

"I'm not asking," Charley said. "The boss is telling. He wants Fred to room here with you until he gets settled. Figures you two boys have a lot of old times to chin about. Keep each other from getting lonely."

"Me . . . take him in here?"

"You want me to tell the boss no?"

"Maybe if he knew . . . ?"

"He knows," Charley said. "That's why he wants it. this is boot camp for you, kid. If you're gonna let personal feelings come ahead of the good of the organization, he wants to know it now. Not too late."

"But throwing that louse in my face?"

"If you can't take that, maybe he won't believe you can take worse."

"I have to do it, I guess. Huh?"

"The boss knows best, Dicky. It's like when you were hiding out. A couple of bad days at first . . . and then it's really over. Done. Believe me, it's the way."

"All right," Link said. "I'll take him in."

"I knew you would," Charley said, his voice hearty and relieved. "I told the boss you were a good boy, Dicky. He thought you'd make trouble with Fred at first. But I told him different. 'Don't you worry about that Dicky,' I said. 'He's a real organization man. No more heart than a pickle. He lives for his work,' I told him. 'Look what a good boy he's been since we took him in. Salutes when you tell him what to do. Clicks his heels. Good troops there,' I told him. 'No cheap hood you can't trust. This boy is loyal. True blue. Full of character, and that's hard to find, character.' Why, I told the boss, 'You tell him the organization needs his right arm, and he'll cut it off for you himself.' Right, Dicky?"

"They took care of me," Link said. "I owe 'em something."

"Sure, Dicky. Be nice to Fred when he comes. No hard feelings."

"No," Link said, wanting Charley to go. "No hard feelings."

Charley went out, and Link sat down. No hard feelings. Come in, Fred, and sleep with me now. And how is my wife these days?

No hard feelings. How dirty could a guy get to feeling? How dirty could a guy get? How'd a guy get so dirty?

He was taking an awful lot from Chicago. More than he had a right to take. They were beginning to push him too hard. Maybe they didn't know it, but Link Aller wasn't a guy who could be pushed all the way around the world without pushing back.

Thinking of how they were making him take Fred in, Link was almost sick with fury. He knew why they were doing it . . . to show him they could do anything they wanted with him and make him do anything they wanted. He'd go along this far, but he had his back to the wall. One more push, and he'd show them . . .

If he only knew what to think about Darlene and himself, it would be easier. But he didn't know what to think, and that made it hard to know what to do or when.

CHAPTER 25

FRED PUT DOWN HIS SUITCASE and stuck out his hand. "Nice place you got here, Link."

"Fred, meet Dick Monroe," Charley said. "Dick, meet Fred Simmons. Shake hands with Fred, Dick."

That was how me he met Fred again. Shaking his hand.

"Now that you two boys have met, how about a little nipper-dipper, Dicky? To cheer us up."

"Yeah," Link said. He exchanged stares with Fred. Fred was looking at his mustache and his clothes. Not the face and clothes he'd expected to see. And Fred had changed—or had been changed. There wasn't much left of the handsome face now. The nose was crooked, and the jaw didn't seem to fit the upper part of the face. As though his face was out of line, with indistinct lines. One cheekbone was sunken in, and the eye above it didn't look like it had been put in right. It was a little sickening to look at.

Link went into the bathroom to pour water for the drinks. How could he do it? How could anybody make him do it? Darlene, the moonlit field, and the rag top. Darlene with ice on my eye where Kern slugged me. Darlene sitting on the six-hundred-dollar couch in the store, holding hands and watching television. Darlene pouring coffee. Darlene and Fred.

He had to repress a crazy giggle. Pouring a drink for Fred. Giving him a place to sleep. While Charley cheered, and the boss cheered, and all the Chicago boys cheered—for what? What kind of life was this, that could make a man pour a drink for the guy? What kind of people cheered? What kind of people were they? Was he? It was crazy, crazy . . .

"Cheers," Fred said, lifting his glass to Link. His new dentures were duller than the old, obviously artificial. He saw Link staring at his mouth. Fred tapped his teeth. "At the rate I lose these things, who can afford two hundred bucks a set?" He laughed and downed his drink.

Crazy people, Link thought. They were crazy, all of them. It wasn't that they were nice guys who stole cars and held up banks and stores for a living. They were crazy. Even Charley, who'd seemed like such a nice guy. There was something about this whole life that was like Fred's face. Twisted and put together crazy.

"We're gonna have a little party for Fred tonight," Charley said. "I guess I'll leave you two alone now. You probably have a lot to talk about. See you both tonight. We're gonna eat and have a little fun."

Charley went out, hiding his crossed fingers. He waited outside the door, listening.

Fred and Link stood quietly. Then Fred sat down. His injured right eyelid had a trick of suddenly dropping closed. The nerves didn't work right anymore.

"Sure is good to be back in Chi' again," Fred said, leaning back.

"I guess so."

"You like it here?"

"It's okay."

"You will when we get busy. I hear we're gonna team up. Maybe with one more boy. We'll give 'em hell, won't we?"

"I guess," Link said.

"You ever pulled a stickup of any kind?"

Link shook his head.

"You've got something coming. You haven't lived until you've done that. Running cars . . . that was a kid's game. I asked to get you for a driver, you know."

"You . . . asked?"

"Yeah." Fred shook his head, sighing. "Remember the night I took you on that Polk City trip with the Caddy?"

Link had to grin a little, despite himself. "You mean the night I took you."

"I was scared pea green," Fred said. "But I figured you knew what you were doing."

"Guess I was scareder than you were, then," Link said. "I knew I didn't know what I was doing."

They both chuckled, looked at each other, and let themselves laugh openly about it.

"I knew right then that I wanted you to be my driver if I got back in my own line," Fred said. "I'm pretty fussy about who I ride with on a job. You can't work with just anybody, you know. You have to find a guy who makes it a team. A guy with guts, timing, a feeling for the job. Like you."

"Let me get you another drink," Link said. "You look like you could use it."

"Yeah, I had a lousy trip up on the train," Fred said. "Had a seat next to an old lady who wanted to tell me all about her diseases."

"Ain't it always the way," Link said without thinking. "Some old guy gets the seat next to the good-looking gal. That's the way it always happened to me when I was going after cars."

There was a momentary, awkward pause.

"Not too much this time," Fred said. "I like a drink, but I'm not a boozer, you know?"

"Sure thing. Real light."

"Remember," Fred said, raising his voice so Link could hear him. "Remember the farmer who thought we were cheating him when we sold him that Merc for three hundred dollars? The one who came back for another one at the same price after he'd had it checked?"

"That one . . . yeah . . ." Link laughed.

Charley uncrossed his fingers and walked away.

— — —

It was a quiet party to begin with. Link, Charley, Fred, and the thin man with the bad teeth, the "middle boss," whose name, Link found out now, was Hamak.

They sat at a round table, each eating according to his fashion. Hamak had stomach trouble and ate oyster stew and crackers. Charley had a steak. Fred had a seafood plate (easy to chew), and Link had spaghetti. Link and Fred sat side by side, Fred's tricky eyelid falling almost every time he leaned over to take a bite, and Link wrinkling his upper lip high on his teeth, to keep the sauce off his mustache.

Hamak looked searchingly at Link and Fred with each spoonful of stew he carried to his lips and sucked in through his teeth. They were talking to each other like old friends.

"See," Charley said in a low voice. "I told you it would work. Not a thing to worry about. He's got the girl out of his head, and he's ready to go to work."

Hamak grunted, unconvinced. "I ain't so sure, Charley. Don't like some of the looks he gives Fred."

"They're like brothers," Charley said. "They ain't even had an argument over socks and ties since they started living together."

"I got to be sure," Hamak said.

"What do you want?"

"Prove he's our boy. Shove it down his throat. Now."

"Okay," Charley said.

Link had turned away from the table and was watching the floor show. Charley leaned toward Fred and whispered in his ear. Fred looked doubtful, then grinned. "All right," he said. "If you'll back me up. He's tough."

"Sure," Charley said. "I'll cheer for you."

Link watched the show, convinced he was having the best time of his life. The new feeling of equality with the others and the new freedom combined to provide a feeling of happiness and contentment. This was the life, all right. Good food, shows to watch, and plenty of excitement. It was the life.

He watched with a strangely nervous interest as a young blonde dancer took the spotlight. Young, blonde, with blue eyes, a delicate, hesitant step, smooth, creamy skin. Her brief costume exposed most of her lithe, beautiful body.

"Whew!" Charley said loudly. "I never saw one like that before."

Fred threw his arm around Link in a comradely manner. "Guess Dicky and me are one up on ol' Charley, eh, Dicky?" he cried. "Guess we know where the twin is, don't we, Dicky?"

Darlene!

Link shoved his chair back and leaned forward, as though to spring at Fred. Link's face was twisted into a dull red mask of hate.

They were watching him. Fred grinning, but with the weak eye fallen shut, leaving the other roundly open and watching. Charley, his teeth clamped down on his cigar, his face beet red, but his eyes staring coldly. Hamak, wary and contemptuous, like a coiled snake.

They knew. They all knew. What was it? What was he mad about? He shook his head to get rid of the confusion, the buzzing, the mixed-up things that didn't make sense. For a second he had been killing mad about something over a girl. For a second he had been somebody else, somewhere.

His mind rocked dizzily, spinning past pictures of a yellow rag top, a blonde girl, a lean mean kid with black hair, sitting under a yard light cleaning spark plugs, some kid in boots and denims getting smashed in the face by a cop, a drugstore, a night in the car in the country, a face he was trying to remember, a person he was trying to be.

The music blared, people laughed, dishes rattled. Outside the grind and growl of heavy traffic, the lonesome sound of too many people in too many cars fighting for too little space with horns and gears.

Dellville, Darlene. Pictures on a postcard, tiny through the wrong end of a telescope. He was one of the pictures. Somewhere the someone he used to be . . . Want that face again, that name again, that girl again, that car again, that town again, that anger again. The other fellow's anger. The other fellow who was himself and was supposed to be mad.

"Eh, Dicky?"

Dicky? He jerked his head erect. They were looking at him. Fred, Charley, Hamak. Sure. Now he knew where he was. Chicago. This was real. Chicago. What was Fred saying? Charley?

He licked his lips uncertainly, feeling scared. He was two people. But who was he? Which one?

"Eh, Dicky?" It was Charley's voice.

"What?" he asked stupidly.

"No hard feelings?" Fred asked, his face twisting into a grin.

They were looking at him. They knew. They'd thrown it right in his face. The kind of thing you killed a man for. The kind of thing Link Aller would have killed a man for. Link Aller didn't take anything from anybody. No matter who they were. He fought. Link Aller fought.

They were looking at him. They knew. They were waiting. They were real. Dick Monroe was real.

Link signaled for a drink. He had always avoided liquor, which had destroyed his father and made his own childhood barren. But in this crisis, when he was forced to face what his life had made of him, it seemed to him a last desperate antidote. When the drink came, he downed it in a single, long gulp, then ordered another. "No," he said thickly, trying to blot out the pain and humiliation. "No hard feelings." He drowned the corpse of Link Aller in the distasteful and unfamiliar liquid.

"You see," Charley said to Hamak. "He's our boy all the way."

Link drank rapidly, to blot out his shame. Not even knowing that he talked, he babbled away at Fred. "I'm gonna kill you someday, Fred," Link said with drunken gravity. "Kill you dead." He fell toward Charley. "Ah'm gonna kill him someday. But don' you worry, Charley. Don' worry. Ah'll wait . . . Ah'll wait . . . Wun hurt ornization fer nothin' . . . Show you . . ." He turned to Fred, trying to see his face, but Fred's face kept slipping and sliding away from him. "You see, Fred," he said, wagging his finger in the air, "I goin' to kill you someday . . . But ah'm good Chicago boy. Take orders . . . Ah'm goin' be your frien' until I goin' to kill you." His head rolled from side to side, as though the words were tipping it. "You mah goo' frien', Fred. You bes' frien' until I goin' kill for . . . for somethin' . . . Sharley tell me, when I gonna kill you. Sharley tells me ev-er-y-thin'. Don' you, Sharley . . . ?"

"Sure thing, Dicky," Charley said. He bit off the end of a fresh cigar. "Satisfied, Al?"

"Yeah," Hamak said, getting up. "I'm satisfied: They'll make a good team, him and Fred. I've got some ideas for this summer for them two. Meanwhile, see that he meets some girls."

Link kept drinking until he suddenly broke out in a cold sweat and felt the world slipping away from him. Fred and Charley helped him to the men's room and pushed him inside. When he felt better, he got to his feet and looked in the mirror over the lavatory. There was a face there, but he was too sick and dizzy to recognize it. It was a pale face, with a black mustache, topped with long black hair. It looked familiar, but when he tried to focus his eyes on any feature, it seemed to waver and dissolve, and it wouldn't hold still and be anybody.

– – –

"Always fussing with the baby," Kern said as he came into the house for lunch.

"He gets hungry," Darlene said. "All the time."

"You ought to let your ma take care of him for a while and get out a little."

"I will," Darlene said. "When I feel like it. Open your mouth, Ricky. Eat . . ."

"De . . . de . . . de . . . baby . . ." Kern said, trying to get the baby to open his mouth. He took out a pair of handcuffs and jangled them over the baby's head. The baby grasped the handcuffs and put one rim in his mouth.

"Pa!" Darlene cried indignantly. "Don't let him put those dirty things in his mouth."

"I've seen worse teething rings," Kern grumbled. "He can't hurt himself on these. No sharp edges or anything."

He chuckled and raised his voice. "Agnes . . . come in here a minute, honey. Want you to see Virgil Kern's grandson cuttin' his teeth on handcuffs. Bet that boy grows up to be a real policeman."

Darlene tried to get a spoonful of cereal into the baby's mouth. "Ain't Grandpa silly," she giggled, opening her mouth when the baby did, and smiling to see what a chore it was for him to learn to swallow.

CHAPTER 26

THE DAY CAME WHEN THEY GAVE HIM A GUN. A .38 automatic complete with shoulder holster. He got into the harness, then put on his coat, self-consciously patting the place where he imagined a bulge would show.

"Perfect!" Charley cried. "And believe me, we don't let you out of the store if it doesn't fit."

"Practice getting at it," Fred advised. "But do it with an empty gun when you're alone."

Link moved his shoulders, trying to get used to the weight of the gun. "What's the celebration?"

"How would you like to be a bank robber?" Charley asked.

"Anything you say," Link said. "What bank?"

"Something small and unguarded," Charley said. "There's no point in picking some big place with guards all over the joint. But there's plenty of small banks that can be had for the asking. And their money's just as good."

"Sounds good to me," Link said. "Got one in mind?"

"Hamak has," Charley said. "He wants to break you in as a driver, and he wants to start you in familiar territory. You're from southern Iowa, aren't you?"

"Yeah."

"What town?"

Link hesitated for a moment. Then, not knowing quite why, he lied. "Chariton." He waited, but Fred didn't say anything. He tried to remember if he had ever told Fred where he really came from.

"Do you know a town called Dellville?"

Link fussed with his shoulder holster so he could avoid looking at Charley or Fred. "I've been there. Usual small town."

"They've got the kind of bank we like," Charley said. "One man and three women, no guards. They're having a carnival of some kind next week. We'll slip down and clean out the bank while the place is full of strange cars."

"Why Dellville?" Link asked.

"Like I said, you know that country, and Hamak wants your first big job to be on roads you know. You know the roads, don't you?"

"I've raced cops on all of them," Link said, "nobody can beat me on those roads. Where do you want out, Kansas City or St. Louis? We can go either way."

"We'll go whichever way they chase us," Charley said.

"Okay."

Charley and Fred began discussing plans for the raid. Link went into the kitchen and ran water into the sink, as though he was getting a drink. He felt shaky and upset. Dellville. In a year he had forgotten the town, forgotten his past, forgotten himself. Now the memories flocked back to stir him, and he felt a strange, fierce longing to be his old self again, with his own name, in his own town. As though he had been lying in an open grave and had suddenly been called back to life.

Charley yelled from the living room. "Come on and listen, Dicky. You have to know this stuff too."

Link joined the two others. Charley was looking at a road map and discussing routes. Fred looked excited and pleased. "It's about time they let me at a bank," he said as Link entered the room. "You watch me work, Dicky. This is my real racket. Boy, it's been a long time."

"Okay," Charley said. "Here's how we move in . . ."

- - -

When they left Des Moines and headed south to Dellville, Link felt that he was home again. It was a road he could almost drive with his eyes closed, he'd been over it so many times at high speed. And now, he was on it again.

Every turn, every flat, every hill flashed a new memory.

The road was haunted by the sound of his old rag top's engine, the squeal of his tires. So many times, the race home.

He drove past the spot where he'd once had a terrific fight with Ricky Madison and whipped the kid in front of his girl. And from that point on, he drove another trip. The last race at night, when Ricky had led him toward Dellville.

His hands trembled as he approached the turn at the top of the hill. It was here, going around this turn and down this hill, that Ricky had floorboarded himself and Sharon to death. Link guided the car around the turn and started down the hill, staring at the bridge at the bottom of the hill. That's where Ricky had hit. That's where he felt a sudden impulse to steer, to follow Ricky.

"How much longer?" Charley asked from the back seat.

Link drove across the bridge. "We're almost there."

"Take it easy. We don't want to get arrested for speeding when we should be arrested for robbing the bank."

"Okay. I'll be careful."

Link glanced in the rearview mirror. Charley's face was redder than usual. Fred was asleep—or seemed to be. Link looked at himself. Besides his mustache, he wore dark glasses, and a straw hat with a bright band. No one would know him. No one would notice, not on a carnival day.

He drove into Dellville, onto the streets of home, and slowly circled the town square. Charley and Fred both examined the town carefully. It was filled with cars, but not many people were on the streets. They could hear the noise from the carnival grounds. The music of the merry-go-round and Ferris wheel, the screams of girls on the thrill rides, the amplified bally.

"Park somewhere," Charley ordered. Link found a spot and nosed the black Ford he was driving into an empty slot. "What now?" he asked.

"I need cigars," Charley said. "Won't have time to get them after work."

"Let me get them for you," Link said. "I need a stretch after the drive."

"Okay. Want to go with him, Fred?"

"No," Fred said. "I'm saving myself for the job."

Link got out and walked toward the Dellville Drug. He knew he was taking a chance at being recognized, but he didn't care. In a way, he found himself almost hoping that he would be seen. He wished Charley and Fred would go away and leave him in Dellville. Now, back in his hometown, only it was real, and Chicago was far away. A name that meant nothing. This was real. This was where he belonged.

He stepped inside the Dellville Drug. There was a girl behind the soda fountain. Exactly where Darlene had been the first night he'd seen her. Darlene. He was disappointed that the girl wasn't Darlene. For a moment he had hoped the clock had turned back, and they were starting all over, again. The second time, he'd know what to do. The second time, he wouldn't get in trouble.

"Can I help you?" the girl asked.

Link pointed to the cigars he wanted. "Six of those."

"Anything else, sir?"

He shook his head. A girl he never saw before. She didn't know him either. He looked around the store. There was a strange boy sitting at the end of the counter, waiting for the girl to rejoin him when she was through with her customer. A boy in boots and denims. Sitting where Link Aller belonged, talking to a girl who should have been Darlene Kern . . . Aller . . . his wife . . .

He put the cigars in his pocket and went toward the back of the store.

"Something else, sir?" the girl asked.

"Use the phone," Link said.

He dropped a coin in the phone and dialed a number he had never forgotten, his heart pounding. He felt unreal, like he was moving through a dream. The phone buzzed in his ear. Once . . . What would he say when she answered? Twice . . . If she was there, what did he want to say? "Hang up!" his brain cried. "Hang up, you fool!"

"Kern speaking."

Link stared in horror at the telephone. "Kern speaking . . ." The hard voice, the tough, official tone. "Hello . . . Hello . . . This is Kern . . ."

In the long moment that followed, Link heard a thin sound of crying in the telephone receiver. He listened to it hungrily, until Kern's voice sounded again, sharp and angry, and Link hung up. He walked toward the front of the drugstore with a fixed, dazed expression on his face. Over the phone, he knew, he had heard for the first time in his life the voice of his baby.

At the other end of the line, Kern shrugged his shoulders and hung up.

"Who was it, Virgil?" Mrs. Kern asked.

"They didn't say. Didn't say a word. Somebody scared to talk when I answered, I guess."

"That's strange."

"Well, it probably don't mean anything. But I think I will take a stroll about town. All these strangers in for the carnival, it don't hurt to keep a watch on what's going on."

– – –

"We get out here and walk to the bank," Charley said to Link. "You wait about half a block back, so nobody will notice you being close to the bank. Keep the motor running like you were waiting for somebody back there. When you see us come out, move fast. You make it to us by the time we're on the street. We jump in, and from then on it's up to you."

"I'm ready," Link said.

"Keep your eyes peeled for us," Fred said. "We'll be moving fast. And you'd better."

"I know what to do," Link said, his fear making him irritable.

"Let's go, if you're ready."

He drove the car slowly along the street until he was about thirty yards from the corner, then stopped, with his engine running, his eyes fixed on the bank door. He saw Fred and Charley walk up the steps and enter the bank. It didn't seem real. The bank was being robbed, he was part of the robbing gang, and it didn't seem real. Not here in Dellville.

His eye was caught by a flash of yellow moving on the far side of the square. His mouth opened and closed soundlessly. It was his old car, with Darrell still driving it. His old rag top! The sound of its pipes twisted his heart as Darrell squirreled around the corners.

Fascinated, Link followed the yellow convertible with his look, aching inside with the hurt of the past. And turning to watch it, he saw two familiar figures in blue trousers and gray shirts. Kern and Arnie VanZuuk, not fifty yards behind him on the sidewalk, talking together. The sight of the two men paralyzed him. He couldn't look away, and he was afraid they would look toward him and recognize him. If only Arnie was alone!

He was snapped back to reality by the sound of a man's hoarse scream. He looked ahead and saw Fred and Charley in the street, wildly looking for him as they each carried a bundle under one arm, and a pistol in the other hand.

Startled, Link threw the car in gear and shot forward, slamming on his brakes as he reached the two men. Shouting curses, Charley yanked open a back door and scrambled in. "Come on, Fred!" he yelled.

At that moment, Mr. Madison, the bank cashier, ran out of the bank with a gun in his hand, and started shooting at the car. Link heard the slugs hit, but he still couldn't believe it was real. He saw Fred crouch and fire back, and Mr. Madison fall back and tumble on the steps. The air seemed filled with the sound of shooting, and when he saw Fred firing toward the rear as he scrambled into the car, Link knew that the policemen were shooting.

Charley and Fred were screaming and raving from the back seat as Link wheeled through the streets and headed for the open country. Cursing him madly because he had delayed. But all he could think of was how Mr. Madison, Ricky's father, looked when he fell.

Numbly driving at top speed, Link felt the most terrible sense of crime and guilt he had ever known. He had forced the son into death. He had brought killers to destroy the father. He had brought death to the streets of his town. He had hurt Dellville, his home.

And from now on, this would be his life. Fear, flight, and blood. Until somewhere, sometime, it caught up with him.

Until a bullet stopped him, and he entered the grave—or prison, if he lived. What would they say, in Dellville, when they discovered that it was Link Aller who had brought home the robbery and the murder? And Link Aller who had carried the robbers and murderers away? That Link Aller was a robber and a murderer in his own hometown?

The chase took shape behind them. Flying over the highway, hating Fred and Charley as they cursed him, he looked in his rearview mirror and saw them come after him. And in the lead, a bright spot of yellow. His own car was chasing him!

He watched with mingled despair and pride as the yellow car moved closer. Oh, how that rag top could move! He had built it. He knew. And now it was chasing him. Running him down.

"Get moving!" Charley bawled from the back seat. "They're gaining."

"I'm flat out now," Link said.

Behind him, Charley and Fred kneeled on the back seat and watched the pursuit, reloading their guns. Link heard the shots as Charley and Fred leaned out of the car and tried a few long ones, to discourage their pursuers. Shooting at his old rag top.

Suddenly it was too much. The violence to his beloved home streets, the chase by his dearly beloved yellow car. His own past life was rising against him in horror. What had he done? What was he doing? With Fred and Charley, until death? Back to Chicago? Back to Hamak's orders? Back to being Dicky Monroe? God in Heaven, what had he done? No more . . . no more . . . no more . . .

The guns behind him exploded and sent bullets whistling back toward his car, his friends, his town. And he was helping the strange murderers escape . . .

They were on a straight, flat stretch of highway. No other cars were coming. It only took a second. He kept the gas pedal to the floor and jammed on his brake, throwing Fred and Charley to the floor. At the same instant he spun the wheel to the left, then let it go back to the right. The Ford began to spin on the dry concrete, and then, let go, it rolled, over and over and over and over again, until its battered hulk tumbled into the ditch with the crash of metal and glass still echoing in the air.

VanZuuk and Kern squatted side by side as they looked at the three inert bodies inside the crushed car.

"Something familiar about the driver," Kern said. "Seems like I've seen him before."

Arnie pushed down for a closer look. "Ah," he said in a sad, husky voice. "Ah, it's too bad. That one we know, Virgil. It's Link Aller."

"My son-in-law," Kern said in a harsh voice. He spat. "I ain't surprised. Once a punk, always a punk."

CHAPTER 27

IN THE HOT EARLY EVENING, Virgil Kern took two chairs from inside the police station and carried them outside. He sat on one and tilted it back against the wall. He took off his cap and wiped his forehead and waited. He loosened his belt a notch and shook his head. He was beginning to get extra inches around his waist. Somehow, he didn't care. He rolled a cigarette and began to smoke. The town was quiet. Almost quiet. A few kids were circling the square in cars, sounding their pipes. Virgil yawned.

In a few minutes, Arnie VanZuuk came into sight, approaching the police station with his slow, rolling walk. Although he was no longer a policeman, he continued to wear his blue pants and gray shirt. At first it had angered Virgil. Now he didn't care. Little by little, he had got used to having Arnie around, giving his slow advice about how to care for the town. And in time, the policeman in Virgil had won out. He accepted the good professional advice, then appreciated it. After a while, it got so he missed Arnie if the old man didn't show up. And so, they had fallen into an informal but definite routine of cooperation and companionship. The way it was, Virgil mused, they were the only two men in town who understood what it was to be a policeman, and he had to talk to somebody who understood.

Arnie reached the police station and dropped into his chair, slowly tilting it back against the wall. He took off his civilian cap and let it rest on his knee. "Hot night," he said to Virgil.

"Sticky," Virgil said. "It might mean rain, if we're lucky."

They sat together, sharing the familiar sounds of their town at night. The familiar, reassuring sounds that signaled all was well. The noise of kids playing the new jukebox in the drugstore, the transient blasts of talk and music from car radios, the dit-dot of unhurrying, honest feet along the sidewalk, the thin echoes of old men arguing under the trees.

"I been thinking, Arnie," Virgil said.

"Yah?"

Virgil pulled a folded page of newspaper from his pocket and read it by the light of the police station window. It was the *Dellville Booster*'s account of the bank robbery. Banker slightly wounded, one bandit dead, another dying, a third would probably live. One of them a local boy.

"That crash," Virgil said. "It don't make sense."

"Something went wrong, maybe," Arnie said. "A bullet."

"Nothing went wrong," Virgil said. "It wasn't a bullet because I was in the lead car, and I wasn't shooting. Doggone it, Arnie, we've both chased Link enough times to know how that boy can drive. And there he was, on the flat, with nothing between him and the border but thin air. I seen it all. All of a sudden them brake lights went on—you seen where his tires drug—and then he spun. And Arnie, that started out like a controlled spin. I could tell that."

"So?" Arnie asked. "What do you think?" He smiled in the darkness.

"I think he wrecked that car a-purpose," Virgil said. "I know he did. But why? He could have got away."

"Maybe he didn't want to get away."

"That's what I figured. But why not?"

"Maybe," Arnie said slowly, "maybe when he seen the town, he liked it here. You know?"

"I never did like him," Virgil said. "But he wasn't no coward. I can say that in his favor."

"He'd been better off if he'd learned to back down once in a while," Arnie said.

"Nobody likes to back down," Virgil said. "Some of us don't know how. But if he done what I think, and spun that car a-purpose, that took an awful lot of guts."

"He would do it if anybody would," Arnie said. "If he had a reason."

Kern shifted uncomfortably. "I can't help but think that anybody with that much guts can't be all bad. Anybody that would kill himself when he could have got away. He didn't know he'd crawl out alive."

"It was a hard way to turn honest," Arnie said.

"What do you think he'll get?" Virgil asked.

"Depends. They could put him away for a long time. Still, if he smashed up on purpose . . . That would earn him time off."

"How do you think folks would feel around here if they knew he done that?"

Arnie shrugged. "I think they would like to see one of our boys get a break. We've known Link so long, some of us."

"Suppose we spoke up for him," Virgil said.

"That would be very nice," Arnie said. "Very nice for him. Maybe a parole to you. But I don't understand . . ."

"The boy's got a wife and baby to support," Virgil said. "I ain't going to feed them forever."

"But I thought . . ."

"So did I," Virgil grumbled. "But the deed's done, and they've got a baby. It might sound funny, but a man likes to see his daughter tie up with a feller that's got some guts. Even if he don't always use good sense. And she seems to like him. Can't everybody marry a millionaire that never had troubles. I was just as wild and ornery as him when I was young, back in Oklahoma. Might have got in the same kind of mess, but for luck. Always lookin' for a fight."

"I never figured to hear you talk like that about Link," Arnie said.

"I never figured to," Virgil said. "It come to me when I was home and seen Darlene gettin' herself and the baby dressed pretty to visit Link in the hospital. It come to me then that the baby with my blood had his blood too, no matter what. And for the first time I realized that Link was family. And a man don't run out on his family just because there's trouble."

"She's seen him?" Arnie asked.

"Every day. The only one they'll let in. You understand what I mean about his being family, don't you?"

"Yah," Arnie said.

"Reckon other folks in town would understand if I stood up for him?"

"There ain't nobody in Dellville who wants to be mean, Virgil. I know this town. It's got room."

Virgil sighed with relief. "I wasn't sure, and I was wondering."

"Sure."

They sat in silence, listening to their town's night sounds.

"I been thinking, Arnie," Virgil said.

"Yah?"

There was a touch of complaint in Virgil's tone. "This here job's getting to be too much for one man. Town this size ought to have two policemen."

"That's right," Arnie said comfortably. "We could use another feller. Some young feller who wants to learn the police work. Give you some time with your family."

"I wasn't thinking of a young feller," Virgil said, slowly building a cigarette. "What's a young feller know about keeping the peace in a small town? Young feller walks around with one hand on his club and the other on his pistol, makin' trouble. I was thinking of an old feller. An old fat one."

"You got an old fat feller in mind?" Arnie asked, touching his shirt at the spot where he had once worn a badge.

"You're always hanging around here anyway," Virgil said gruffly. "You might as well git paid for it."

"But the city council. They might not . . ."

"I already told them what to do. They're gonna vote it at the next meeting."

"That's a nice thing you done, Virgil," Arnie said after a moment. "I got a couple good years to give. When I ain't no good anymore, I take off the badge myself."

"We'll worry about that when we come to it," Virgil said.

"Maybe," Arnie said, "I stay until you find a good young feller. If there was some young feller who knew the town, and they knew him . . . Somebody who might understand what it's like to be in trouble . . . especially the kids . . ." Arnie risked a glance at Virgil.

"I was thinkin' that too," Virgil said. "I don't know if we can do it or not. I'll have to find out. Have to talk it up around town, too. I could train that boy right, Arnie. Him and me could make a good team for this town in a couple years. Work together until little Ricky is maybe big enough to join his pa. If it works out. I surely would like that, Arnie. I always did hope I'd have a policeman for a son-in-law. Him and me, we could really look after this town. If it works out."

Virgil took off his police cap and put it under his chair. He unbuckled his gun belt and pulled it away from his body, rubbing his sides where the belt had chafed. He tucked his feet under his chair and hooked his heels over the lowest rung.

Virgil closed his eyes and listened to the sounds of his town. The music, the voices, the footsteps, the old men in loud argument under the trees, the restless mutter of dual exhausts as the boys circled the square, Arnie's slow, heavy breathing, the nervous shiver of leaves against leaves as the trees passed along the rumor of rain. "Yes sir," Virgil said, "this job is gettin' to be too much for one man."

Arnie nodded in the darkness, contentedly fanning his face his hat. "By golly," he said. "It sure is a nice night."

ABOUT THE HOT ROD SERIES

Originally published in the 1950s, this series of six popular rodding books sold more than eight million copies. Immensely popular with young readers in its early days, the books are wildly entertaining cautionary tales meant to keep speed-hungry teens safe on the streets. In limited circulation for the past decade, Octane Press has created this fantastic new edition.

The Hot Rod Series by Henry Gregor Felsen includes:

HOT ROD
ISBN 978-1-64234-089-1

STREET ROD
ISBN 978-1-64234-104-1

CRASH CLUB
ISBN 978-1-64234-131-7

ROAD ROCKET
ISBN 978-1-64234-132-4

FEVER HEAT
ISBN 978-1-64234-133-1

RAG TOP
(originally published as CUP OF FURY)
ISBN 978-1-64234-134-8

www.ingramcontent.com/pod-product-compliance
Lightning Source LLC
Chambersburg PA
CBHW020859180526
45163CB00007B/2562